Model Science Teacher Preparation Programs

Model Science Teacher Preparation Programs

An International Comparison of What Works

edited by

Jon E. Pedersen
University of South Carolina

Tetsuo Isozaki
Hiroshima University

Toshihde Hirano
Aichi University of Education

IAP

INFORMATION AGE PUBLISHING, INC.
Charlotte, NC • www.infoagepub.com

Library of Congress Cataloging-in-Publication Data

A CIP record for this book is available from the Library of Congress
http://www.loc.gov

ISBN: 978-1-68123-800-5 (Paperback)
978-1-68123-801-2 (Hardcover)
978-1-68123-802-9 (ebook)

Printed and bound by CPI Group (UK) Ltd, Croydon, CR0 4YY

CONTENTS

INTRODUCTION

Jon E. Pedersen
University of South Carolina

Tetsuo Isozaki
Hiroshima University

Toshihide Hirano
Aichi University of Education

In recent times much has been written and communicated both in the popular press and in research-oriented publications about the performance of students on international math and science assessments. Many countries are held as exemplars not only in the performance of their students but also how teachers are prepared to teach their students. Undeniably, there is clear evidence that teachers have a great deal of impact on student performance yet little has been done from a comprehensive point of view to examine the requirements (national, regional, local), processes, and procedures used in the preparation of science teachers. It is with this in mind that Dr. Tetsuo Isozaki in December of 2014, brought together in Japan a group of international scholars from 5 countries: Finland, France, United Kingdom, United States, and Japan. The purpose of this meeting was to discuss and share promising practices for the preparation of science teachers. It struck us that there is much to learn from each other regarding the challenges

Model Science Teacher Preparation Programs, pages vii–x

and opportunities associated with science teacher education. Questions, challenges, and opportunities that, for the most part, each country has addressed in some manner, but overall a synthesis of these practices has not been collected in a single volume as we have attempted here. We recognize that we cannot include every country that prepares science teachers in a volume such as this. Therefore, we have spent considerable time looking at international data to determine a list of countries that could be included in this volume and realize that some may disagree with our assessment of countries that have been selected. We believe that our selection is a worthy representative sample of countries that have performed well on international assessments and have other accessible data that would reflect a high level of performance for students and science teachers. We also wanted to insure global representation. We did not want a "Eurocentric" or western hemisphere view only. We sought and invited individuals from both the eastern and western hemisphere as well as countries representing the diverse areas of the world. We started with a list of over 25 countries and individual scholars that we invited. Of those, roughly 18 responded in some way to our request. Ultimately, 14 countries accepted our offer to contribute a chapter and are represented in this volume sharing the challenges, opportunities, and promising practices as it relates to the preparation of science teachers.

Although we fully realize the arguments (pro and con) for such assessments, we believe that this provides some basis for the effectiveness of science teachers in a country and ultimately (although indirectly) the quality of a science teacher preparation program. The limitations to our method of sampling are many and we would not suggest that this volume represents the "best" or the only preparation programs that should be highlighted. We are suggesting, however, that each country represented herein has something significant to contribute to the discussion of science teacher education. And, that we can all benefit and learn from each other and continue to strive to improve the quality of science teacher education in each of our own respective countries, universities, and programs. The countries included in this volume are as follows:

Argentina	Japan
Australia	Korea
Canada	People Republic of China
Finland	Portugal
France	Turkey
Germany	United Kingdom
Israel	United States of America

(alphabetical order)

This volume, *Model Science Teacher Preparation Programs: An International Comparison of What Works* draws on the expertise of each author(s) to create a much-needed comparison of science teacher preparation from around the world. This book takes direct aim at this gap in knowledge and will provide the opportunity for readers to make an in-depth analysis of each countries' science teacher preparation requirements as well as how each country and various universities prepare science teachers. Within this book you will find that authors are *not* just reporting on the success of their nation and/or program (although this is a key component of each chapter). Rather, we have asked authors to take a critical look at the process by which science teachers are educated and share with the reader both the positive aspects of what is happening as well as the challenges each country faces in science teacher preparation. As you will see, the following questions were used as a guiding framework for each chapter:

1. What is the structure of the national requirements for science teacher preparation and what is the history (brief) behind these requirements?
2. What, if any, are the accreditation or national assessments (tests) required of countries or programs for science teacher preparation (e.g., tests to qualify for certification or licensure, tests to qualify for jobs, etc.) and how has this impacted science teacher preparation (both positive and negative impacts)?
3. What are the standards for each country regarding science teacher education (national, state, regional, prefecture, etc.) and how has this impacted science teacher preparation (both positive and negative impacts)?
4. Are there different levels of certification or licensure within your country, how are these determined and how has this impacted science teacher preparation (both positive and negative impacts)?
5. What requirements are "universal" for each university that prepares science teachers and how has this been impacted by: national requirements, accreditation, national assessments, standards, and levels of certification?
6. For universities in general, what are the content requirements (number of hours, number of courses, number of units, etc.) for discipline or content area license (e.g., certification or licensure for a physics teacher, biology teacher, etc.)?
7. What are the general pedagogical courses required for science teacher preparation (e.g., learning theory, special education, reading in the content area, etc.)?
8. What are the content pedagogical courses required for science teacher preparation and are these content specific (i.e., chemistry,

physics, life science, earth science) or are these general courses covering all content areas?

9. What are the specific field experience requirements (how many hours in the classroom and specifically what role does your science education student play) both officially (required by university, state, prefecture, national entity) and customary (but not part of a mandate) for your science teacher education students and when do these experiences occur (e.g., early in the education process, after students graduate and are certified or licensed: apprenticeships)?
10. An analysis of the strengths and or areas for improvements of science teacher preparation in your country including (but not limited to):
    a. What content knowledge is critical to know and understand to be a science teacher (chemistry, physics, life science, earth science)? Here we are not expecting a long laundry list of concepts or specific "facts" but a general sense of what you recommend as necessary to be successful in the science classroom. You may also refer to specific documents (e.g., national requirements, professional association requirements, etc.).
    b. What general pedagogical knowledge is critical to know and understand to be a science teacher (e.g., learning theory, assessment, classroom management, etc.)? Here we are not expecting a long laundry list of concepts or specific "facts" but a general sense of what you recommend as necessary to be successful in the science classroom. You may also refer to specific documents (e.g., national requirements, professional association requirements, etc.).
    c. What pedagogical content knowledge is critical to know and understand to be a science teacher? Here we are not expecting a long laundry list of concepts or specific "facts" but a general sense of what you recommend as necessary to be successful in the science classroom. You may also refer to specific documents (e.g., national requirements, professional association requirements, etc.).
    d. As you will see, each author(s) has addressed these questions in various ways. Additionally, the final chapter of the book is an analysis of all chapters submitted for the book. This chapter focuses on what works, what is promising, and the obstacles we face as science educators as we prepare science teachers around the globe.

CHAPTER 1

SCIENCE TEACHER EDUCATION

A Multi-Country Comparison—United Kingdom

Paul Davies
Institute of Education, UCL

Ruth Amos
Institute of Education, UCL

This chapter explores the provision of science teacher education within the United Kingdom. In our exploration, we review the history of this provision, the current situation with teacher education and training, and critically discuss the implications and future of how science teachers are prepared in the United Kingdom. The United Kingdom is a country formed from England, Scotland, Wales, and Northern Ireland and currently has a population of around 62 million. There is neither an over-arching education, nor one teacher education, system within the United Kingdom; the separate countries have levels of organization, both local and national, which

Model Science Teacher Preparation Programs, pages 1–27

are specific to each context. This chapter mainly focuses on the situation in England, which has the largest population, is the seat of the central U.K. government, and plays host to the full range of science teacher education programs.

THE STRUCTURE OF THE REQUIREMENTS FOR SECONDARY SCIENCE TEACHER PREPARATION

In understanding the current and potential future state of science teacher education, it is important to have an appreciation of the history of, and complex approaches towards, teacher preparation in England. Knowing something of this history helps not only to rationalize current models of teacher education but also reveals the multi-faceted relationships between schools, government policy, universities, and other teacher education establishments in a climate of socio-political change.

The History of Science Teacher Education in England

The history of science teacher preparation in England, in terms of policy, theory, and practice is a complex and, as a result of continual governmental interest in educational reform, at times turbulent one. This has by no means ended; at the present time, we are in a state of major change, which will have potentially long-lasting effects on the landscape of how we prepare teachers at the start of their training and throughout their careers. For example, the Welsh Assembly recently published the Furlong (2015) review into teacher training provision in Wales, which makes a strong case for research-informed courses. In England, the Carter review (2015) has resulted in initial teacher training continuing to make a somewhat unsettling transition from primarily university-led programs/curriculum towards more school-led provision.

Science as a school subject has special status as a "core" subject along with English and mathematics. This means it is mandatory for students to study science in some form from when they first enter school, at age four or five years, until the age of 16.[1] As a core subject, there has been considerable interest in how science teachers are prepared, the form and shape of science teacher education, ideas about the relationship between theory and practice and the purposes of science teaching in school. A detailed history of teacher education in England can be found in Robinson (2006), in which she explores the key ideas of the establishment of formal teacher education and its change over the past two hundred years against the historical contexts of school and social reform.

Robinson (2006) identifies two inter-connected components central to an understanding of approaches to education, not just with regard to science teachers, but all teachers. The first is the dominance at different times of either school-based or apprenticeship-type approaches, in contrast with university and other higher education institutions (HEIs)-based models of teacher education. Broadly speaking, in the nineteenth century, science teacher education was located within schools, with teachers training "on the job," with little in the way of "formal" learning taking place beforehand and almost no opportunity for recognizing links between theory and practice (for a detailed account of this period see, for example, Jones, [1924]). The second, is the relationship that teacher education institutions have with schools and the changes that we observe happening here. This became most evident when secondary school became compulsory for all in the mid-twentieth century, but we still see the complexity of this relationship evolve today as the social and political context of education in the United Kingdom continues to shift, which we explore in part later on.

Before science teacher education was formalized in the United Kingdom in the early nineteenth century, teachers of the upper and middle classes tended to be graduates of either Oxford or Cambridge University,[2] often with clerical backgrounds. The onset of compulsory schooling meant that almost overnight thousands of additional children suddenly required teachers, and it is here that a model of school-based, albeit brief, teacher education was introduced. This shift in the model of teacher education is important because, as is discussed below, teacher education is currently revisiting the apprenticeship model, which for some has echoes of the historical class-divided, elitist ethos. Throughout the nineteenth century, the model of in-school training grew but was controlled in an *ad hoc* way with schools recruiting novice teachers into the profession on a needs basis. This changed in 1902 when the Education Act established new local education authorities, (LEAs) which took over responsibility for teacher education. This was coupled with a shift to college-based elements of the training and the provision of training bursaries and other types of financial support. As this new approach to teacher education became more established, university-based programs were opened up in an attempt to attract more graduates into the profession, mostly destined for teaching in secondary schools, with college-based training providing routes into primary school.

The Robbins report (1963) on higher education marked a major reform in this provision and saw the establishment of Bachelor of Education (BEd) degrees and Post Graduate Certificates of Education (PGCE; Postgraduate Diplomas in Education, PGDE in Scotland), meaning that teaching became, for the first time, an all-graduate profession. By the early 1980s, teacher education moved towards being under tighter government control with the birth of two government agencies responsible for regulating teacher education: the Teacher Development Agency (TDA) and the Office for Standards in Education (Ofsted: a non-ministerial government department with the role of monitoring education. One of their roles is in inspecting schools by observing lessons and examining documentation to assess educational standards). This development in government control was coupled with a shift in university-based teacher education towards a practice-based model, with greater emphasis on in-school training. This resulted in a reduction in exploration of the relationship between theory and practice. It was also at this time that flexibility of training options increased, with a range of employment-based routes and flexible PGCE models opening up.

School Types in England

Over the last ten years, there has been a major change in the classification of school types in England. Here we outline the most important aspects of these changes, for a comprehensive account, see Department for Education (DfE, 2015a) and Richmond (2013). Until 2002, there were broadly three types of schools in England: state maintained, grammar, and independent. State maintained, or "comprehensive" schools are funded through local government controlled agencies that allocate funds and oversee spending. Grammar schools, of which there are now very few in England, are a selective type of state maintained school, with free-at-the-point of entry education. Unlike comprehensive state maintained schools, grammar schools require students to sit an entrance examination, often a national standard examination, called the 11+ (as it is taken at age 11 years old). In this sense, grammar schools are selective, granting admission to those children who perform best. Independent or "private" schools are non-state maintained schools, often established through charitable donations, which charge fees at the point of entry. These schools almost always have entrance examinations, similar to the 11+ examination, although students may also be admitted with scholarships awarded for other strengths such as sporting or musical ability. The fees charged by private schools are often extremely high; those offering boarding facilities charge thousands of pounds per year. Historically, there has been a perceived elitism associated with these three school types, which means independent schools are often deemed the most prestigious.

In 2002, the Education Act, drawn up by the then Labour government, allowed a new type of school to be established known as an academy. Academy schools are state-funded but receive money directly from the government education department, rather than the local education authority, giving them much greater freedom in terms of its deployment. Also, unlike state maintained schools, they have much greater freedom in terms of the taught curriculum, design of school structure, and teacher pay scales. There are broadly three types of academy. The first, sponsored academies, are state maintained schools which have been forcibly "transformed" into academies by the government (usually following an inspection by Ofsted which identified severe weaknesses). These are "sponsored" or overseen by government-approved groups, which can be other schools, charitable organizations, or special interest groups. The second, converter academy schools, are those which made the choice to become academies; these do not need to be managed by government-approved sponsors. The third, free schools, are new academy schools formed since 2011 that are funded directly from the government and fairly autonomous in terms of curriculum design. Free schools can be established by groups, for example, parents and teachers, in response to perceived local community needs, who apply to the Secretary of State for Education to make the case for setting them up.

These changes in school classification have implications for teacher education because it means schools have, in some cases, much greater autonomy in how they are organized, how teaching and learning takes place, and the potential for academy chains to develop a corporate ethos which may have implications for how trainee and newly qualified teachers (NQTs) feel about working in them. The new academies have won praise in some areas, with some "failing" schools being seen to have "turned around" very quickly once they have been converted into an academy. One of the most successful and famous of these is Mossbourne Community Academy, a school in East London that before becoming an academy had poor examination results (Tomlinson, 1997). Sir Clive Bourne, a local philanthropist, donated large sums of money in order to rebuild the school, and a new "super head teacher," Michael Wilshaw, was brought in. These modifications, together with a major overhaul of staff and changes to approaches to teaching and learning, resulted in rapid progress of the students and greater overall intellectual success at the school (Fowler, 2011). Moreover, Mossbourne is only one of many schools that have undergone the conversion to academy status.

Yet, criticism has stemmed from trade unions, teachers, parents, and politicians who have cited problems such as the expense of re-designing/building schools, the fact that academies tend to expel a greater number of children than local state maintained schools, (Parkinson, 2005) and the role that private sector entrepreneurs now play in publicly-funded education. Especially relevant to science teacher education, and a good example of some of the criticism leveled at academies, has been a number of sponsors who have associations and affiliations with religious groups setting up new academies. While religious schools are not uncommon in England, their status is controversial (Johnson, 2013), and some of the religious affiliations now supporting academies are extreme in nature. The most famous example is possibly the Emmanuel Foundation School group academy chain founded by Sir Peter Vardy, who has been accused of promoting creationism in the schools (Guardian, 2009). In 2002, the head of Science at one of the schools was alleged to have given a lecture to students and parents promoting creationist thinking. Following the controversy around this, he resigned from his post. Since then, the schools have been under greater scrutiny both from Ofsted and the media, resulting in mixed reports. Some suggest that the provision of education at the school is excellent (Wainwright, 2006), but others argue it promotes creationist thinking (BCSE, 2006; Pike 2009). Whilst there is no published material on how science teachers approach working in these schools, one might imagine that the schools attract teachers with certain religious views and perspectives on science that may be contrary to established scientific ideas and explanations. Teachers working in such environments may also find it hard to challenge these views and feel intimidated, making their own professional development problematic and ultimately affecting the quality of teaching and learning.

The Question of School Versus HEI Teacher Preparation

Science teacher education, and all other teacher education for that matter, stands at a crossroads in England. Over the past ten years, successive governments have sought to move teacher education further and further towards school-based experience, meaning that HEIs have no, or minimal, influence on teacher preparation. This move was initiated by Kenneth Clarke, when, in 1992, he gave a speech as Secretary of State for Education, in which he stated that teacher education, or, as described then, teacher training, should be 80% school-based. In addition, the schools involved in this training should be government-selected and subject to government control through the Ofsted inspection framework (e.g., see Ofsted 1996). These changes meant that schools would work with HEIs, share funding for courses, and have greater influence on the preparation that the trainees received. At the same time, to aid clarity, the HEIs involved were renamed as initial teacher training (ITT[3]) providers, and they effectively had to "bid" to work with schools in developing ITT-school partnerships. This bidding process has led to a marketplace where ITT providers and schools jostle to provide, and achieve, both value for money and quality training. This is by no means straightforward, and the tensions and challenges faced at the time of this reform are still very much visible today.

While few would doubt the need to have strong and cooperative relationships between ITT providers and schools, not to mention the necessity for trainee teachers to spend much of their training in school-based environments, the way this relationship was established in the 1990s has left teacher education with a legacy of top-down government control and little effective consultation taking place between government departments, schools, and ITT providers. In 1996, Furlong and Smith argued that a disregard for the importance of HEIs in providing trainee teachers with a solid grounding in educational theory, along with a shift towards a highly skills-based framework for teacher education, was a dangerous move that would have long term, negative consequences (Furlong & Smith, 2013). It was also around this time that a key player in the ITT provider-school partnership was formalized: the school-based subject mentor, a teacher who should receive training from the ITT provider designed to support the trainee teacher during school-based training periods. As Husbands (2001) argues, the importance of the role that the mentor plays as a conduit through which the ITT provider's "vision" and the technical aspects of being a teacher are realized cannot be underestimated. However, this relationship is under threat both in terms of time for release for professional development of mentors, and particularly in areas in which high teacher turnover in some schools results in a loss of continuity in mentoring.

REGULATION OF THE KNOWLEDGE AND SKILLS THAT TRAINEE TEACHERS REQUIRE

The establishment of Ofsted and the TTA in the 1990s marked a major change in the requirements for ITT courses, with the government having much greater control over prescribing the content and structure of training curricula and pathways (Richards, Simco, & Twistelton, 1998). A major concern at the time was that ITT providers needed to have a standardized program and that they should be held accountable in terms of educating and producing high quality teachers. This shift towards a general greater scrutiny of public sector services was accompanied by the beginnings of Ofsted inspections and a report on the quality of training provided at all levels of ITT. Over the next five years, regulation and standardization became more formalized until, in 1997, the TTA developed and implemented a set of mandatory standards (known as the Teaching Standards) against which all trainee teachers would be measured (see the Department of Education and Employment [DfEE], 1997). The new standards were generic in nature and aimed at all trainee teachers, covering six aspects of teacher knowledge and skills: subject knowledge, planning, teaching and class management, monitoring assessment, recording, and reporting, together with professional issues, which was a lengthy list of professional behaviors deemed appropriate for teachers (see DfEE, 1998). The Teacher Standards were used to assess each trainee at the end of their initial teacher training period, grading them 1–4 ("very good" down to "unsatisfactory").

The Teachers' Standards have remained an important aspect of the regulation of ITT providers and have undergone revisions over the past 15 years, particularly in response to political imperatives requiring focus on specific national priorities in school (for example, behavior management and literacy development). However, the adjustments have evoked some criticism, which we will discuss with particular reference to science teachers.

CURRENT MODELS OF SCIENCE TEACHER EDUCATION IN ENGLAND

As a brief exploration of the history of teacher education in England has revealed, there have been shifts in the relationship between ITT providers and schools. A core principle that has remained is the need for trainee teachers to follow a training program which leads to the award of qualified teacher status (QTS). The regulation of this is now controlled by the National College of Teacher and Leadership (NCTL), a government agency responsible for developing strategies to improve education. QTS is a qualification broadly recognized across the world as confirming for all parties that an appropriate level of competence in teaching has been achieved. QTS is a requirement for all teachers in state-maintained schools in England and Wales (with similar accreditation occurring in Scotland and Northern Ireland, under different names) although teachers in independent schools are exempt, as have been those in academies since 2012.

During the last five years, we have seen a major shift in training provision, with school-based training routes being favored by the conservative-liberal democrat coalition (elected to government in 2010), and now the conservative government (elected in 2015). In the current model, trainee teachers can opt for a training program that is managed and run by an HEI (university-based route) or by a school (school-based route). The school-based routes are varied in terms of the relationship they may or may not have with an HEI and also the control that the school has over recruitment and design of the training program.

Routes Leading to QTS

There are currently six routes leading to the award of QTS in England and similar awards in Wales, Scotland, and Northern Ireland. *Bachelor* degrees are undergraduate degrees, normally 3–4 years in length and involve studying for a major in a subject discipline and a minor in education.

Students spend time at both university and at school and graduate with either a Bachelor of Education (now more common for primary level education), a BA with QTS, or a BSc with QTS. The popularity and provision of bachelor level degrees leading to QTS has shifted since their inception in 1968, and currently new BSc programs leading to QTS are being developed in the physical sciences, as one way of addressing a short-fall in teachers with physics backgrounds (DfE, 2015b). Despite this move, most training routes leading to QTS are at postgraduate level.

The vast majority of teachers train via the *Post Graduate Certification in Education (PGCE)* route. This course is normally one year in duration and involves the trainee teacher attending both an HEI for lectures, workshops and tutorials, and undertaking school-based experience (typically known as "school placements"). The PGCE is not a postgraduate degree, but rather an advanced qualification at level 7 leading directly to a vocational career. Since the period between 2005–2007, most PGCE courses moved towards offering courses leading to the award of masters-level credits, which contribute to a full master's degree at a later stage. A small number of providers still offer courses at higher- (H-) level (NB trainees qualifying at H-level would be awarded a PGCE, indicating below masters-level study, known as the Professional Certificate in Education). To gain a PGCE and QTS, trainees must typically pass two M- (or H-) level modules and a practical teaching module.

The next three routes share a common approach in that training is led by schools. *School Direct Non-Salaried (SDnS)* is the most recent addition to available training routes. Schools are responsible for recruitment but have to work with an accredited provider (either a university or a SCITT). Providers and schools work closely together, often following the PGCE model. *School Centred Initial Teacher Training* (*SCITT*) routes, only available in England, are based entirely in school. Some SCITT programs lead to the award of PGCE as well as QTS, whilst others do not. The *School Direct Salaried* (*SDS*) route is a graduate, school-based training program in which trainees work as unqualified teachers in school, receiving a salary and learning "on the job" through an apprenticeship model. Again, schools must work with an accredited provider.

The final route is *Teach First*. Established in 2001, this is a charitable organization aimed at training high-quality graduates to teach children in low-income communities in England and Wales. A two-year, salaried postgraduate course, it follows a school-based approach, alongside association with an HEI leading to the award of PGCE and an optional Masters degree in leadership. The program is designed to attract the "best" graduates (with at least a higher second class degree) and is highly competitive. The name *Teach First* hints at the design of the program as one in which graduates teach for two to three years, before moving into other forms of employment. Smart, Hutchings, Maylor, Mendick, and Menter (2009) and others have argued that retention of teachers following this route is lower than other routes. Moreover, the training is more expensive and the program is not cost effective. However, there is a growing number of Teach First Ambassadors (teachers from the program) who are now supporting new recruits in challenging schools, which has the potential to stabilize the teaching workforce to an extent. As is explored below, each of these routes has its own benefits and challenges, and it is important to consider these against a backdrop of major change in teacher education in England.

All of the teacher education programs described above associated with HEIs now require the payment of tuition fees. Currently, these stand at a maximum of £9,000. Government loans are available to cover the fees and living expenses, but this means that trainees are normally left with debts at the end of their studies, which can be considerable if they have also completed a loan-supported undergraduate degree (Bachan, 2014). Graduates opting for say a Teach First training route, for example, often do so as they need to earn whilst training.

Requirements for Entry to a Science Teacher Education Program

All teacher training programs that lead to QTS for science teachers specify minimum entry requirements. Irrespective of the program, in England all candidates must complete professional skill tests (known as QTS Skills Tests). These are designed to ensure all trainees have appropriate numeracy and literacy skills and are used to "assess the core skills that teachers need to fulfill their professional role in schools, rather than the subject knowledge needed for teaching" (DfE, 2015c, para. 1). The tests are computer-based and a pass mark of 63% is stipulated. They cover basic numeracy and literacy and also test usage of technology skills through the online interface. Candidates may take the test a maximum of three times before they can be accepted into a training course; failure on the third attempt causes them to be barred from taking the tests for two years, effectively meaning they are unable to train to become a teacher for three years. No such tests exist in the other countries of the United Kingdom, and the (2015) has recommended that the need for such tests be re-examined.

Beyond the skills tests, all candidates must have a grade of C (perceived as a "good" grade) in English and mathematics in the General Certificate of Secondary Education (GCSE, a national examination which children take at age 16 years) or equivalent. If candidates wish to study for an education degree at undergraduate level, (i.e., BEd, BSc with QTS) they must have obtained passes at least at D grade in two post-16 level science qualifications (the most common being advanced level examinations typically gained at age 18–19).

Entry into all postgraduate programs requires a first degree in a science subject at least to the level of "third class" (the level above "pass"). In reality, very few candidates enter a teacher training program with a degree class below "lower second" as a result of funding implications (as noted above, Teach First candidates must have achieved at least a "higher second class" degree). Where candidates might fall below this "degree standard" they might be required to complete a subject knowledge enhancement course (SKE course) to support the development of key school-subject knowledge. These courses are funded by the government and are available in teacher shortage subjects (e.g., physics and modern foreign languages; see Get Into Teaching, n.d.). In addition, in order to help boost applications for postgraduate science teacher education programs, successive governments have provided incentives in the form of scholarships. These are awarded to all trainee science teachers enrolled in a recognized program that leads to QTS and their value varies from £9,000 to £25,000, depending upon the science specialist subject and the class of degree the candidate obtained. Currently, physics graduates are awarded the largest scholarships due to a shortage of physics teachers. The different requirements for entry also, inevitably, mean that candidates and employers have certain perceptions about the "quality" of teachers from different programs, especially when linked to student achievement in school (Harris & Sass, 2011). These perceptions are compounded by Ofsted reports generated through inspection of accredited ITT providers (inspections normally take place around every three years). Grading that comes about through this process means competition to be accepted onto a particular course can be intense; however, applications from suitable candidates wanting to become physics teachers still remain low for all ITT providers.

It is not unusual for ITT providers to accept candidates onto their programs with science qualifications above undergraduate level, for example, masters or doctoral. In addition, a significant number of candidates have considerable experience in employment, sometimes in fields unrelated to science. This presents some challenges, as "career-change teachers" may find it difficult to reconcile their perceptions of good teaching and learning with those of ITT providers and schools (Tigchelaar, Vermunt, & Brouwer, 2012).

Requirements of Science Teacher Training Programs in the United Kingdom

As discussed above, all ITT providers in England are subject to regulation by the government's DfE, which acts directly through the NCTL, TDA, and Ofsted. In order to be awarded QTS, all trainee teachers must show they have met a benchmark of competences set out in the government-produced standards, the Teachers' Standards (DfE, 2012). Part of the statutory requirement associated with "meeting the Teachers' Standards" decides the amount of time that trainee teachers must spend in schools. As

TABLE 1.1 Required Number of Secondary Subject Training Days in School in England

Course Program	Days (Weeks) in School
4 year undergraduate	160 (32)
1, 2, or 3 year undergraduate	120 (24)
Secondary graduate (non-salaried, e.g., PGCE or SD tuition)	120 (24)
Secondary graduate (employment-based salaried, e.g., SD)	as determined by the program

Table 1.1 shows, this varies across programs but starts at a minimum of 120 days in two different schools; there are implications for trainees' experience due to these differences:

Since 2012, the Teachers' Standards (DfE, 2015d) have been presented as a series of knowledge, skills, and attributes that all teachers must demonstrate within two sections (Part One and Part Two). Guidance stipulates that the contents of all ITT courses should be designed to ensure that trainee teachers meet all of the relevant teachers' standards for the age range in which they are attempting to qualify. Therefore, there is an implicit requirement to prepare trainee science teachers to teach the content of the English National Curriculum (NC) for Science (DfE, 2015e). The content of the NC, thus, by pragmatic necessity dictates the units of study in taught sessions at university. Some examples of curriculum content are illustrated later on.

Part One of the Teachers' Standards is concerned with teaching, namely: setting high expectations for students, promoting student progress, demonstrating good subject curriculum knowledge, planning well structured lessons, promoting inclusion in the classroom, using assessment well, managing behavior, and supporting the wider life of school. Part Two is concerned with personal and professional conduct, namely: treating students with dignity, promoting safeguarding in school, showing tolerance, and promoting British values. This final standard has attracted some concerns, associated with the definition of "British-ness." As Beauchamp, Clarke, Hulme, and Murray (2015) point out, in England the shift towards a more "craft-based" approach to teaching and learning and a move away from the involvement in HEIs is problematic in terms of the Teachers' Standards because it means schools may promote their own local versions of how competence should be demonstrated. This is particularly the case for those standards that are not clearly defined (Goepel, 2012). With regards to British values, the government has been challenged to provide clear definitions and to explain how one might be measured against them. Within the Teachers' Standards, they are explained as: "democracy, the rule of law, individual liberty and mutual respect, and tolerance of those with different faiths and beliefs" (DfE, 2015d, p. 14).

The particular challenge comes with how these values are reconciled within an association with the United Kingdom's current PREVENT strategy, which emerged partly in response to the government's anti-terrorism policy (Heath-Kelly, 2013; TSO, 2011). Part of this strategy requires all teachers to be vigilant and responsive to signs of radicalization and extremism within schools. It also includes being aware of safeguarding issues associated with cultural practices, for example, female genital mutilation. While, arguably, most teachers would not question the relevance of much of what they are expected to be mindful of in school, the guidelines on how to detect and respond to the government requirements to, for example, "identify extremism in the classroom," is unclear and could potentially be concerning for many teachers. The concept of teachers "spying" on their students is also an uncomfortable prospect for many, given that teachers always strive to encourage positive views about respect and democracy through their teaching and professionalism (Baker-Beall, Heath-Kelly, & Jarvis, 2014).

Beyond provision designed to allow trainees to meet the Teachers' Standards, there is no prescribed curriculum for teacher education courses, but ITT providers are expected to demonstrate how they assure consistency of assessment in reaching the standards. This is measured and checked through the Ofsted inspection framework, but again this can be problematic. A major criticism of the process has been that the inspection does not focus on the teaching and learning that the trainee teacher experiences through HEI-based training but has come to be almost solely associated with teaching experiences in school. Information about the provision of training from HEIs is obtained through scrutiny of documentation during inspection and testimony by trainee teachers, obtained through interviews between the trainee and inspectors. The dissociation of inspection between HEI providers and the school-based experience of ITT is perceived by some as a definite move by the government to argue for an increase in school-only based ITT routes (Simmons & Walker, 2013). Surprisingly, this shift in terms of the "quality" and relevance of teacher education is contrary to claims that successive governments have made about wanting to increase the quality of teacher education, for example, through encouraging study at masters level, both in terms of boosting the quality of teacher preparation and raising the profile of the teaching profession (Burton & Goodman, 2011). Only time will tell how this tension is to be reconciled as school-based, "craft-focused" ITT begins to dominate in England.

We will now explore the models and curricula of the key routes in science teacher education in England.

CONTENT REQUIREMENTS FOR ACCREDITED INITIAL TEACHER EDUCATION COURSE IN ENGLAND

The 2014 NC for science in England (DfE, 2015e) is designed for students to learn specified content ideas from biology, chemistry, physics, and earth sciences as well as generic scientific skills in a strand currently described as "working scientifically." In addition, the examination awarding bodies in England interpret the NC for 14–16 year-olds and the core contents for 16–19 year-olds in their own syllabi, under statutory guidance from the Office of Qualification and Examination Regulation (Ofqual). The taught curriculum in schools, and by pragmatic association in trainee teacher programs, is therefore heavily influenced by the need to prepare students for high-stakes assessments.

The Teachers' Standards (Dfe, 2015d) specify the need for development of competent specialist subject knowledge, so biology, chemistry, physics, and physics with mathematics trainees primarily train to teach their own specialist subject(s) to 11–19 year-olds. Examination of the *Times Educational Supplement* (TES) over time, a newspaper where almost all schools advertise teaching vacancies, reveals that many secondary schools in England actually wish to employ specialist science teachers who can teach across the full range of sciences especially at lower secondary level (11–16 year olds). As training programs are developed in partnership with schools, university providers work to ensure that these needs continue to be met. However, this reality can cause confusion and dissatisfaction for some trainees who want to focus fully on their own specialization.

As there are no further mandatory requirements, the necessity for ITT providers to give trainees full opportunity to meet the Teachers' Standards in their specialist disciplines means that universities have devised unique but similar programs for accreditation. For example, typically the one-year PGCE and SD (non-salaried) courses run for 36 weeks and trainees spend 60 days (12 weeks) on the associated university-based components of the program.

On employment-based routes, trainees spend the majority of the training (over the entire school year) working directly in school, so these routes give slightly longer applied apprenticeship than the non-employment routes. Trainees then typically spend only five subject specific and 11 generic pedagogy days in the university, whilst also, with some providers, working towards one masters-level module in the first training year and a second in the NQT year. There is no guarantee that the second training year will take place in the same school as the first year.

The Teach First and university-supported SCITT programs begin with an intensive training period of typically six weeks and usually run in the summer (known as the Summer Institute). The idea is to front-load training before trainees undertake their main school experience. The duration of the two-year Teach First course is, therefore, the whole school year plus the six-week induction in the first year, followed by the whole of the second year in the original school as an NQT. Trainees spend only six days with the accredited provider on developing specific subject knowledge pedagogy. They work towards 60 masters-level credits during the first year and have the option of completing a full masters in the second year.

In the university-led components of teacher training, subject specialists and generic tutor program leaders and tutors have some autonomy in devising module content and units within modules. Once university validation of courses has been achieved, providers have to seek and maintain accreditation from the DfE. During program inspections by Ofsted, provision must be judged as at least "good," the mid-category in the range of quality of provision designated from "outstanding" to "requires improvement." As a consequence, inspection criteria used by Ofsted during a given academic year have a bearing on the content and focus of a training program. What appears to be reasonably open accreditation guidance has to be interpreted through an inspection criteria lens, which is strongly linked to the imperative for schools to focus on high-stakes assessment. Nonetheless, universities are committed to leading inspiring programs which give trainees opportunities to explore and understand how to support learning in creative ways.

General Pedagogical Courses

Trainee science teachers in England develop subject specific and generic pedagogical knowledge together during the taught program in university. Generic pedagogical content is typically initiated through keynote lectures given by experts in the field from the university, as well as external educational strategy, policy, and practice experts including head teachers. Ideas and strategies for building effective practice may then be elaborated upon during mixed subject or within subject-specific workshops and tutorials led by university tutors and science teachers from partnership schools. There are, of course, many complex factors that affect students' learning, so generic strategies are then contextualized during experiences in school through sessions on pedagogy that have a particular local significance. For example, a school may have a large number of immigrant students who are early English language learners, so a school-based session on how to support their learning in science (and other subjects) will be held. Seminars in school are normally led by senior teaching staff and specialist teachers or educational consultants. Science trainees may work with trainees in other secondary subjects as well as with NQTs. This opportunity has the potential to create supportive communities of early career learners. Without such support and without compassionate and evaluative support from their subject (and other) colleagues, young teachers may leave the profession relatively quickly (School Teachers' Review Body, 2014). A typical sequence of generic pedagogical topics is shown in Table 1.2.

There are a number of critical factors that support science trainees making progress with notions about what makes for effective learning and teaching. Ideas, concepts, and practice initiated at the university need to be built upon and enriched throughout the two school experiences. Trainees then have the impetus, and modeling by experienced mentors, to make links between content (subject matter) knowledge and the pedagogical approaches that support student's learning. We are convinced that both experiences in university and in schools are critical for this connection to be successful. At university, there is a strong emphasis on learning through reflection-in-action (Schön, 1983). Trainees read associated core texts, as well as professional and academic journal articles to support their understanding and to link theory with practice. Typical core readings are shown in Table 1.3.

TABLE 1.2 Typical Sequence of Generic Pedagogical Topics Followed by Secondary Trainee Teachers[a]

University-Based Generic Sessions (Module 1)	School-Specific Generic Sessions (School 1)
• Teaching in schools • Inclusion in the secondary classroom (respecting difference) • Working with early English language learners • Teachers' legal responsibilities • Being a professional teacher • Introduction to classroom management • Evidence-based teaching • What are British values? • Dealing with issues of forced marriage and gender-based violence • Voice care for teachers	• The school as an organization • Planning and assessment • English as an additional language: subject specific language and learning strategies • Safeguarding: policy and practice • Assessment for learning • Behavior for learning • Inclusion: Special needs • Inclusion: equal opportunities • Being a form tutor (pastoral care) and working with parents • ICT and computing across the curriculum
University-Based Generic Sessions (Module 2)	**School-Specific Generic Sessions (School 2)**
• Applying for your first teaching post • Adolescents and adolescence • Securing achievement for all • Inclusion: understanding autism • What do we mean by education? • Teachers as researchers • Re-visiting behavior for learning	• Re-visiting planning and assessment • Inclusion: working with bilingual students • Working with other professionals • Re-visiting behavior for learning • Primary-secondary school transfer • Citizenship and Personal, Social and Health Education (PSHE) • Inclusion: raising standards for all • Post-16 education

[a] These may vary to some extent across providers, based upon local needs and preferences.

TABLE 1.3 Typical Core Texts and Journal Articles of Postgraduate Initial Science Teacher Education Programs

Generic Texts and Articles	Science Texts and Articles
Brooks, V., Abbott, I., & Huddleston, P. (2012). *Preparing to Teach in the Secondary School (3rd ed.).* Maidenhead, England: Open University Press. Capel, S., Leask, M., & Turner, T. (2013). *Learning to Teach in the Secondary School (6th ed.).* London, England: Routledge. Kyriacou, C. (2007). *Effective teaching in schools: Theory and practice (3rd ed.).* Cheltenham, England: Stanley Thornes.	Toplis, R. (2015). *Learning to Teach Science in the Secondary School (4th ed.).* Abingdon, England: Routledge. Wellington, J., & Ireson, G. (2012). *Science Learning, Science Teaching (3rd ed.).* London, England: Routledge. Abrahams, I., & Millar, R. (2008). Does practical work really work? A study of the effectiveness of practical work as a teaching and learning method in school science. *International Journal of Science Education, 30*(14), 1945–1969. Black, P., & Harrison, C. (2004). *Science inside the black box: Assessment for learning in the science classroom.* London, England: Nelson.

CONTENT PEDAGOGICAL COURSES REQUIRED FOR CROSS-SCIENCE SUBJECTS

Science subject specialist course units are guided by the English NC for Science (DfE, 2015e) in order to cover the secondary age ranges 11–19 in biology, chemistry, and physics. Physics with mathematics trainee teachers focus on both physics and mathematics curricula (typically mathematics for 11–14 year-olds and physics for 14–19 year-olds). Subject knowledge content continues to be a rather narrow, traditional series of topics and ideas that have been present in the science NC since its inception. For example, in biology 11–14 year-old students learn about the structure and function of living organisms, material cycles, energy, genetics, and evolution. In chemistry, topics include the nature of matter, the periodic table, the Earth, and its atmosphere, while in physics, students explore energy, motion, forces, and waves (DfE, 2015e). A fourth strand now known as "working scientifically," focuses on the scientific method including asking scientific questions that can be investigated, planned, carried out and allow for data measurement and analysis. The science curriculum now suggests that applications and implications of science should be studied as part of the wider curriculum. However, we suggest that science teachers need to take full part in giving students opportunities to consider global and social issues in which science plays a part. In other words, scientific literacy should be a key goal of the science curriculum.

What is far more interesting than the specific content is two-fold: how graduate scientists transform their own expert subject knowledge for *school science* and how they learn the pedagogical approaches that will best support students' learning in school. A typical subject pedagogical session at the university allows the trainee to map the important ideas and skills that students should learn from ages 11–19 (see Table 1.4). There is then a focus on the 14–19 age range to facilitate depth of understanding in subject specialization.

TABLE 1.4 Typical Sessions Taught in a Teacher Science Education Program

Age of Student (Curriculum Stage)	Chemical/Physical Idea	Typical Learning Activity	Learning and Teaching Strategy
11 (KS3)	Matter can be solid, liquid, or gas	Observing objects, identifying characteristics and properties of matter	Practical work Discussion Inquiry
11 (KS3)	Matter is made of tiny particles called atoms	Magical scissors: What would you have in the end if you could keep on cutting and cutting up a piece of paper?	Imagination Modeling Discussion
12 (KS3)	Atoms can join together to form molecules or giant lattices	Role play in which children act as atoms	Modeling Discussion
13 (KS3)	Elements can react to form new substances called compounds	Burning magnesium	Practical work Modeling Discussion
14 (KS4)	Atoms consist of a nucleus and orbiting electrons in shells	Using a simulation	Modeling Discussion
15 (KS4)	Atoms are conserved in a chemical reaction, never destroyed	Predicting and measuring yields of products Using a simulation for balancing equations	Practical work Modeling Discussion Inquiry
16 (KS4)	Atoms have unique mass	Using a simulation	Practical work Modeling Discussion
17 (KS5)	Electrons occupy sub-shells	Using a simulation	Modeling Discussion
18–19 (KS5)	Colors in organic dyes and transition element compounds arise due to electron movements between energy levels	Role playing d-d electron transitions	Modeling Discussion

Take for example learning about the nature of the atom which is one of the "big ideas" in chemistry and physics. If you ask a graduate chemist or physicist to explain the nature of the atom, what would she say (or possibly do)? The chemist may focus on the nature of the electrons on the atom, being perhaps more interested in their influence during chemical reactions, and might draw upon quantum theory and Schrödinger's equation. The physicist may enthral you with notions of quarks, leptons, and neutrinos. However you look at it, aspects of this content (subject) knowledge are far too sophisticated for school science, and the trainee science teacher needs to spiral downwards towards the embryonic ideas which lead there, dust them off, and start again! And even before the age of 11, what are children thinking about the nature of the atom? So, to support the trainee's learning journey, a subject-specific university course provided them with tools to construct a linking framework for subject knowledge and effective pedagogy. Concept mapping forms an important part of this process. The pedagogical approaches for teaching the required ideas and skills are then layered on. For learning about the nature of the atom, such an approach might look something like what's shown in Figure 1.1.

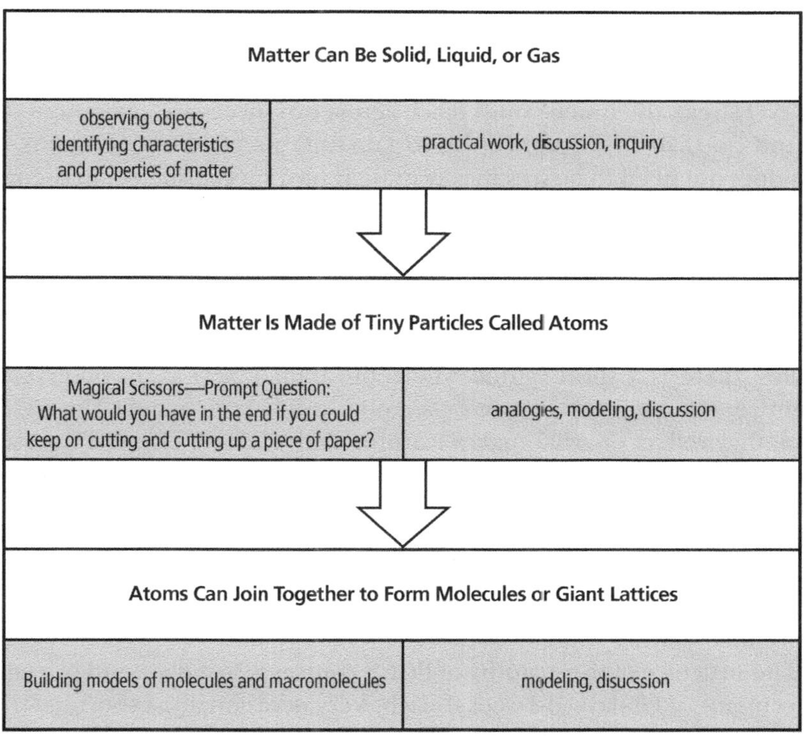

Figure 1.1 Teaching ideas through strategic activities linked to effective pedagogical approaches for learning ideas in science.

So "knowing your stuff" as a science teacher is far more complex than being able to give your own scientific explanation for phenomena. You need to be an excellent, responsive communicator, able to gauge the progress of your students through various means, and be reflective in evaluating your own practice.

FIELD EXPERIENCE REQUIREMENTS

The practical field teaching requirement for becoming a fully qualified teacher in England spans two years. Independent of the route of the first year of training, the trainee spends their second year fully employed in a school as an NQT. An NQT has a slightly reduced teaching load of 20–21 hours per week in England (a fully qualified secondary teacher with no additional responsibilities teaches about 22 hours per week). Teacher training in England is therefore a two-phase apprenticeship; as discussed, the first year may be part-university led and school-based, or school-based and university-accredited.

There are some differences in the school experience pattern across universities and between training routes. The expectation is that science subject specialists will teach their specialism to 14–16 year-olds (or 14–19 with some providers) in at least one school experience during the first year. In the NQT year, the trainee must teach across two successive age ranges (11–14 and 14–16 or 14–16 and 16–19) to gain fully qualified teacher status, but this does not need to be specifically in their own specialism. So, in theory, a biology specialist may fully qualify at the end of the second year by teaching only chemistry and physics in their NQT year. In practice, it is very likely that the specialist will teach at least some of their own subject.

The expected roles for trainee science teachers in their first year differ between training routes. At the beginning of non-employment-based routes, there is a short period where the trainee acts as an observer-apprentice, watching experienced classroom teachers in action. During this time, they reflect on what makes learning and teaching effective in discussion with the teachers and their mentor. They may support small groups of learners and begin to plan and co-teach lessons. They then take on the role of teacher-apprentice as they start to plan and teach their own lessons. In employment-based routes, there is a much shorter period (often very little) of observation or team-teaching at the beginning in favor of teacher-apprentice starting straight away.

The structure of the majority of PGCE courses is based around two equal placements of 60 days (12 weeks) each. Very occasionally, a shorter, "trial" first placement is used, typically 20 days (4 weeks), such that the second placement lasts longer (100 days, 20 weeks) so that trainees work on continuity. Some trainees report that there is too little time to build significant teaching confidence in short first placements, whilst others find this initial "taste" of teaching and school life to be very productive. They then welcome a longer second school experience to develop more in-depth learning relationships with students and with colleagues in the second school.

SD (non-salaried) trainees usually have two school experiences of similar length, but an alternative pattern is an ABA school experience. The trainee has a standard first school experience (A_1, 60 days) followed by a shorter second school experience (B, 30 days), and then return to their first school for a concluding experience (A_2, 30 days). Trainees on the ABA pattern often face challenges in making progress with putting ideas about continuity and student progress into practice if they are training in two very different school settings.

Trainees on the School Direct Salaried program may have been previously employed by the school in which they train in the first year, but, as mentioned earlier, there is no guarantee that their training will continue in the same school for the second year.

Teach First trainees are employed by the same school for the two year training period; they experience a very short second school placement in first year: only five days in a second school. As a result of the time split between university and school on the different routes, science trainees will physically spend different amounts of time in the classroom on practical teaching during the first training year, see Table 1.5.

In the second year of training, all NQTs regardless of chosen route teach on their own (usually no co-teaching is possible), and out-of-class mentoring is put in place. In-class observations and support are organized by individual schools and departments; there is statutory guidance for NQT induction set out by the DfE (2015f). Head teachers have statutory responsibility and typically appoint a senior teacher-mentor in the school to assume responsibility for NQTs (not necessarily a science teacher). Pedagogical training is organized by the senior mentor and may take place in collaboration with other schools. The NQT will typically have six observed lessons spread throughout the year, and judgements will be made on the quality of teaching using Ofsted inspection criteria.

ANALYSIS OF STRENGTHS AND AREAS FOR IMPROVEMENT OF SCIENCE TEACHER PREPARATION IN THE UNITED KINGDOM

In critically analyzing the preparation of science teachers in the United Kingdom, and, particularly England, there are three main areas we suggest emerge showcasing good practice, but which also reveal some of the challenges we are currently facing. These areas are the nature of the science

TABLE 1.5 Route Through Postgraduate Teacher Education Programs

Training Route	1st School Experience	2nd School Experience
PGCE/SD (tuition) 2 equal length school experiences	• 12 weeks duration • Initial phase of observing lessons (2–3 weeks) • Paced introduction of planning and teaching own classes with supporting teacher present (over 2–3 weeks from 1 up to 10–11 hours teaching per week) • Planning and teaching 10–11 hours (supporting teacher observing/giving feedback)	• 12 weeks duration • 1 week observation • Planning and teaching 10–11 hours per week
SD ABA school experience pattern	• 12 weeks duration • As above in school A	• 6 weeks duration in school B: – 1 week observation – Planning and teaching 10–11 hours per week • 6 weeks duration in school A: – Planning and teaching 10–11 hours per week
SD salaried	• Majority of whole school academic year (minus 16 days in university plus up to 6 weeks in 2nd school) • In main training school: – Some initial observation then planning and teaching builds up progressing over the year: recommended 12, 15, 18, 20 hours per week	Typically recommended 6 weeks duration (but varies) In subsidiary school: Working with different classes, some teaching, some support
Teach First	• Whole school academic year plus typically six weeks "summer school beforehand which includes in-school observation • In main training school: – Straight into planning and teaching 15–16 hours per week	Typically 1 week (5 days) duration In subsidiary school: Planning and teaching 11 hours per week
SCITTs	Whole school academic year and starting the previous year (minus subject specific days in the university)	Depends upon the individual SCITT but typically 1 week (5 days) duration

Note: The short "mini" A placement pattern is not included here.

curriculum, the move towards greater school-controlled ITE, and the length of postgraduate science teacher education.

As there is limited learning and teaching time at university, program tutors have to make decisions about which ideas and skills to focus on in taught sessions, which then serve as models for learning and teaching approaches. The content of the English NC for science (DfE, 2015e) is substantial and as a consequence, trainee science teachers need to study during personal time and in their school placements, continually developing their subject knowledge for school science. There are common core ideas and skills in university-led sessions with an important focus on the "big ideas in science" which guides the focus. There will be some variation across providers, governed by factors such as geographical area, local or cultural needs, and educational opportunities often beyond the classroom. We would argue for a balance between pragmatically-chosen topics in response to the needs of curriculum and those which develop enthusiastic, inspiring science teachers. To dispense entirely with the pragmatic themes would be ill-advised.

However, knowing *what* to teach is only part of what it means to be a teacher; understanding *how* to teach is also vitally important. Using Shulman's (1986) notions of pedagogical knowledge (PK) and pedagogical content knowledge (PCK), we recognize that ITT programs make provision for development of these ideas. For example, programs include general PK such as methods of assessment and development of complex explanations, as well as specific PCK (something the teachers' standards, in their current form, do little to support from a subject-specific point of view), building conceptual understanding or developing cross-science links. Early career teachers need to focus on how children learn consistently and HEIs are in the best position to provide access to vital, contemporary evidence-based practice (see for example, Hattie & Yates, 2014). However, as we see the erosion of the role of HEIs in teacher education, we feel this will have a detrimental effect on teachers' expertise as a result of a dislocation between educational theory, which is rich and varied in both education in general, and specifically in science education. We risk trainees losing access to knowledge and skills essential for becoming a rounded science teacher, in terms of recognizing the broader possibilities for a science curriculum such as inquiry, creativity, specialist skills like dissection, and context-based approaches to learning. A good example of this current change in curriculum focus in England is that the requirement to undertake practical work may become diminished through it being removed from high-stakes assessment. So, while the science curriculum in its current form does little to highlight these valuable ideas, skills, and ways of thinking, their importance and relationship to science teaching and learning can be "kept alive" through HEI provision.

We agree with Osborne's (2015) position in which they put forward a science curriculum based upon ways or types of reasoning (thinking). The subject content of such a curriculum could then provide the fascinating ideas and contexts that promote students' curiosity, and university-based components of science teacher training programs could focus more fully on scientific thinking skills.

In conclusion, the history of teacher education in England has seen shifts between school-based and school-HEI based ITT, or, as we would rather ITE. But we now stand at a point where the government is promoting a "school only" model of ITT, with the position of HEI-led ITT being diminished to that of quality assurance and possibly supporting attainment of masters-level assignments. The rise of SCITT and other providers is, to us, of concern as it means the breadth and continuity of national science teacher preparation is lost, as schools take a more "local" approach and an apprenticeship, craft-based model towards producing teachers suited only to their specific context. This shift away from HEI-school partnership also presents issues in terms of how subject-specialist mentors are trained and supported, and leads to the development of multiple, bespoke training programs which, as we are already seeing, offer less than a holistic approach to teacher education. For example, the pattern of school-based experience that SCITTs are permitted to provide is varied, meaning in some SCITTs a trainee may experience as little as five days "teaching" in a different school. The "frontloaded" nature of the training provided through summer schools is also problematic, as trainees have little time or capacity to reflect on their own learning through reflective evaluation of practice. We strongly argue for a return to a strengthened model of HEI-school partnership, which provides opportunities for trainee teachers to make links between theory and practice and reflect on their on-going development.

Nonetheless, the move towards more school-led training does have a key advantage over other postgraduate routes in that it allows the training period to cover both the trainee and the NQT year. At the moment, only the Teach First and some SCITT programs run this way, with minimal HEI involvement. We argue that such a model of teacher education is to be applauded. It is something we would like to see happening on all postgraduate routes because an HEI-school partnership version of this model would strengthen early career teachers' evidence-based practice. A same-school two-year training model affords a trainee time to develop as a teacher, build capacity in personal learning and embed oneself in school culture. This then positions the early career teacher well for making a commitment to lifelong-learning. Sadly, the political will at the moment in England may mean we wait a long time for this to be recognized.

NOTES

1. The current model of learning science until 16 years is under review with the potential for science to be a core subject for students up to 18 years old, these changes sit within wider policy changes about compulsory education.

2. Oxford and Cambridge universities (known collectively as "Oxbridge") are the oldest and most prestigious of the U.K. universities, often being perceived as elitist.
3. The term ITT is now widely used in the United Kingdom to encompass all training routes which lead to accreditation for teaching in pre-university education institutes (schools, colleges, etc.). We concur with Ellis and McNicholl (2015) in arguing for the use of the term "initial teacher education (ITE)" as this reflects the scholarly approach we feel is key to effective teacher learning.

REFERENCES

Abrahams, I., & Millar, R. (2008). Does practical work really work? A study of the effectiveness of practical work as a teaching and learning method in school science. *International Journal of Science Education, 30*(14), 1945–1969.

Bachan, R. (2014). Students' expectations of debt in UK higher education. *Studies in Higher Education, 39*(5), 848–873.

Baker-Beall, C., Heath-Kelly, C., & Jarvis, L. (Eds.). (2014). *Counter-radicalisation: Critical Perspectives.* London, England: Routledge.

Beauchamp, G., Clarke, L., Hulme, M., & Murray, J. (2015). Teacher education in the United Kingdom post devolution: Convergences and divergences. *Oxford Review of Education, 41*(2), 154–170.

Black, P., & Harrison, C. (2004). *Science inside the black box: Assessment for learning in the science classroom.* London, England: Nelson.

British Centre for Science Education [BCSE]. (2006). *Interviews with former vardy schools students.* Retrieved from http://www.bcseweb.org.uk/index.php/Main/ESFInterviews

Brooks, V., Abbott, I., & Huddleston, P. (2012). *Preparing to Teach in the Secondary School (3rd ed.).* Maidenhead, England: Open University Press.

Burton, D., & Goodman, R. (2011). The masters in teaching and learning: A revolution in teacher education or a bright light quickly extinguished? *Journal of education for teaching, 37*(1), 51–61.

Capel, S., Leask, M., & Turner, T. (2013). *Learning to Teach in the Secondary School (6th ed.).* London, England: Routledge.

Carter, A. (2015, January). *Carter review of initial teacher training ITT.* Retrieved from https://www.gov.uk/government/uploads/system/uploads/attachment_data/file/399957/Carter_Review.pdf

Clarke, K. (1992, January 4). *Speech to the north of England conference.* Manchester, England.

Department for Education [DfE]. (2012). *Teachers' standards.* Retrieved from https://www.gov.uk/government/uploads/system/uploads/attachment_data/file/283566/Teachers_standard_information.pdf

Department for Education [DfE]. (2015a). *School leaving age.* Retrieved from https://www.gov.uk/know-when-you-can-leave-school

Department for Education [DfE]. (2015b). *Major push to get more maths and physics teachers into our classrooms.* Retrieved from https://www.gov.uk/government/news/major-push-to-get-more-maths-and-physics-teachers-into-our-classrooms

Department for Education [DfE]. (2015c). *Professional skills tests.* Retrieved from http://sta.education.gov.uk

Department for Education [DfE]. (2015d). *The teachers' standards.* Retrieved from https://www.gov.uk/government/uploads/system/uploads/attachment_data/file/301107/Teachers__Standards.pdf

Department for Education [DfE]. (2015e). *National curriculum: Science programme of Study.* Retrieved from https://www.gov.uk/government/publications/national-curriculum-in-england-science-programmemes-of-study/national-curriculum-in-england-science-programmemes-of-study

Department for Education [DfE]. (2015f). *Initial teacher training criteria: Statutory guidance for initial teacher training providers in England.* Retrieved fromhttps://www.gov.uk/government/uploads/system/uploads/attachment_data/file/434608/ITT_criteria.pdf

Department for Education and Employment [DfEE]. (1997). *Teaching: High status, high standards (Circular 10/97),* London, England: DfEE.

Department for Education and Employment. (1998). *Standards for the award of qualified teacher status (Circular 4/98),* London, England: DfEE.

Education Act. (2002). *Education Act 2002.* Retrieved fromhttp://www.legislation.gov.uk/ ukpga/2002/32/contents

Ellis, V., & McNicholl, J. (2015). *Transforming teacher education: Reconfiguring the academic work.* London, England: Bloomsbury.

Fowler, R. (2011, February 23). *Mossbourne Academy: A tale of high expectations . . . and no excuses. The Telegraph.* Retrieved from http://www.telegraph.co.uk/education/8341428/Mossbourne-Academy-A-tale-of-high-expectations.-.-.-and-no-excuses.html

Furlong, J. (2015). *Teaching tomorrow's teachers: Options for the future of initial teacher education in Wales.* Oxford, England: Oxford University Press.

Furlong, J., & Smith, R. (2013). *The role of higher education in initial teacher training.* London, England: Routledge.

Get Into Teaching. (n.d.). Retrieved from https://getintoteaching.education.gov.uk/subject-knowledge-enhancement-ske-courses

Goepel, J. (2012). Upholding public trust: An examination of teacher professionalism and the use of teachers' standards in England. *Teacher Development, 16*(4), 489–505.

Harris, D. N., & Sass, T. R. (2011). Teacher training, teacher quality, and student achievement. *Journal of public economics, 95*(7), 798–812.

Hattie, J., & Yates, G. (2014). *Visible learning and the science of how we learn.* Abingdon, Oxon: Routledge.

Heath-Kelly, C. (2013). Counter-terrorism and the counterfactual: Producing the "radicalisation" discourse and the UK PREVENT Strategy. *The British Journal of Politics & International Relations, 15*(3), 394–415.

Husbands, C. (2001) Change Management in Education. In R. McBride (Ed.), *Teacher Education Policy* (pp. 7–21). Lewes, England: Falmer Press.

Johnson, H. (2013). *Reflecting on faith schools: A contemporary project and practice in a multi-cultural society.* London, England: Routledge.

Jones, L. G. E. (1924). *The training of teachers in England and Wales: A Critical Survey by Lance GE Jones.* Oxford, England: Oxford University Press.

Kyriacou, C. (2007). *Effective teaching in schools: Theory and practice (3rd ed.).* Cheltenham, England: Stanley Thornes.

Ofsted/TTA. (1996, November). *Framework for the assessment of quality and standards in initial teacher training 1996/7.* London, England: OFSTED.

Osborne, J. (2015). Practical work in science: Misunderstood and badly used? *School Science Review, 96*(357), 16–24.

Parkinson, J. (2005, March 17). *Why the fuss over city academies?* Retrieved from http://news.bbc.co.uk/1/hi/education/4357383.stm

Pike, M. A. (2009). The Emmanuel schools foundation: Sponsoring and leading transformation at England's most improved academy. *Management in Education, 23*(3), 139–143.

Richards, C., Simco, N., & Twistelton, S. (1998). *Primary teacher education: High status? High standards?* Lewes, England: Falmer Press.

Richmond, W. K. (2013). *Education in Britain since 1944.* London, England: Routledge.

Robbins Report. (1963). *Higher education report of the committee appointed by the prime minister under the chairmanship of Lord Robbins.* London, England: Her Majesty's Stationery Office.

Robinson, W. (2006). Teacher training in England and Wales: Past, present and future perspectives. *Education Research and Perspectives, 33*(2), 19.

Schön, D. A. (1983). *The reflective practitioner: How professionals think in action* (Vol. 5126). London, England: Basic books.

School Teachers' Review Body. (2014). *Twenty-fourth report.* London, England: Office of Manpower Economics. Retrieved from https://www.gov.uk/government/uploads/system/uploads/attachment_data/file/318574/STRB_24th_Report_Cm_8886_web_accessible.pdf (Last accessed 26th November 2015).

Shulman, L. (1986). Those who understand: Knowledge growth in teaching. *Educational Researcher, 15*(1), 4–14.

Simmons, R., & Walker, M. (2013). A comparative study of awarding organisation and HEI initial teacher training programmes for the lifelong learning sector in England. *Professional Development in Education, 39*(3), 352–368.

Smart, S., Hutchings, M., Maylor, U., Mendick, H., & Menter, I. (2009). Processes of middle-class reproduction in a graduate employment scheme. *Journal of education and work, 22*(1), 35–53.

The Guardian. (2009). *What a creation.* Retrieved from https://www.theguardian.com/books/2005/jan/15/features.politics

Tigchelaar, A., Vermunt, J. D., & Brouwer, N. (2012). Patterns of development in second-career teachers' conceptions of learning and teaching. *Teaching and Teacher Education, 28*(8), 1163–1174.

Tomlinson, S. (1997). Sociological perspectives on failing schools. *International Studies in Sociology of Education, 7*(1), 81–98.

Toplis, R. (2015). *Learning to Teach Science in the Secondary School (4th ed.).* Abingdon, England: Routledge.

TSO. (2011). *Prevent Strategy 2011.* Retrieved fromhttps://www.gov.uk/government/publications/prevent-strategy-2011.

Wainwright, M. (2006, December 5). They aren't faith schools and they don't select. *The Guardian.* Retrieved from http://www.theguardian.com/education/2006/dec/05/newschools.faithschools

Wellington, J., & Ireson, G. (2012). *Science Learning, Science Teaching (3rd ed.).* London, England: Routledge.

CHAPTER 2

SCIENCE TEACHER PREPARATION IN GERMANY

Knut Neumann
Leibniz-Institute for Science and Mathematics Education (IPN)

Hendrik Härtig
Universität Duisburg-Essen

Ute Harms
Leibniz-Institute for Science and Mathematics Education (IPN)

Ilka Parchmann
Leibniz-Institute for Science and Mathematics Education (IPN)

The structure of the German teacher preparation system is closely linked to the development of formal schooling in Germany. There is not one single German school system, but many slightly different ones. Germany is a federation of 16 federal states, the so called *Bundesländer* or *Länder.* While the federation in this form only exists since the re-unification of West and East Germany in 1990, its origins date back to the foundation of a national state from a number of smaller, independent territories in 1871. At that time, most of these territories had already developed (at least some kind

Model Science Teacher Preparation Programs, pages 29–52

of) a school system. Throughout history, the territories have retained independence in questions of educational politics from the national or federal government, respectively. Today, the Länder still have cultural sovereignty, which leads to the fact that Germany has 16 similar, yet slightly different school and teacher preparation systems and an even larger variety of (science) teacher education programs. In the following sections, we will outline the major steps in the historical development of formal schooling in the German territories or Länder, respectively, in order to identify the main features that are common across the school and teacher preparation systems from all 16 Länder. Following this, we turn to a description of science teacher preparation and selected science teacher education programs.

HISTORICAL DEVELOPMENT OF THE GERMAN SCHOOL SYSTEM

Schooling has a long-standing history in the German territories. Graf and Vogelbacher (2001) describe the historical development as follows: Up until the 6th century, schooling was mostly privately organized and driven by the Greek-Roman idea of education through engaging in the liberal arts. From the 6th century on, an increasing number of monastery schools emerged, the main purpose of which were to educate monks and scholars about religious traditions. From the 9th to the 12th century, a literal schooling boom was to be observed. The number of schools rapidly increased, with the intention of each parish having at least one school. Schools were no longer exclusively affiliated with monasteries; instead, a growing number of town schools emerged. Education in both types of schools was still best characterized as scholarly education through engagement with the liberal arts. As a preparation for education in scholarly schools, elementary schools were set up to teach students the required reading and writing skills together with some basic arithmetic. In the following centuries, scholarly schools opened up more and more to laypeople. Thus, elementary schools became an important source for education in reading, writing, and arithmetic, particularly for those seeking employment in business or trades (Graf & Vogelbacher, 2001, pp. 1142–1143; see also Döbert, 2007).

For a long time, elementary and the various forms of scholarly schools dominated the school landscape in Germany (Sandfuchs, 2004). The *Generallandschulreglement* edict from 1763 strengthened the elementary schools in Prussia: one of the larger and most influential German territories (Graf & Vogelsbacher, 2001). As the edict envisioned formal education for every citizen, this led to the widespread establishment of elementary schools, designed to educate all children from age 5 to age 13, the so called *Volksschulen*: a development that quickly extended to other territories. With the development of a broader middle class in the 18th century, another new school form emerged: the *Realschule* (Döbert, 2007, p. 299). Since education at elementary schools focused on providing citizens with the most fundamental skills needed, the Realschule offered education for those seeking extended education, but who did not intend to pursue an academic career (Döbert, 2007). Finally, with the introduction of the *Abiturprüfung*, that is the general qualification for university entrance, in 1788, a third form of schools emerged from the various forms of scholarly schools: the *Gymnasium* (Graf & Vogelsbacher, 2001). The Gymnasium retained the idea of a scholarly education in the fine arts and was the school form that led to an academic career.

This three-track system soon built the common core of the school systems in different territories or Länder, respectively. Until recently, it persisted with only minor changes. One difference, for example, was the addition of comprehensive schools, the so-called *Gesamtschulen*, as a fourth school type, incorporating all three tracks in some Länder from the early 1960s onward. Other changes included the establishment of a four-year elementary school, the *Grundschule* in the 1920s. All students attended this school before moving on to either Volkschule, Realschule, or Gymnasium. Further, the Volksschule was renamed into *Hauptschule* in the 1960s (Döbert, 2007). Following publications on the findings from the Third International Mathematics and Science Study (Baumert et al., 1997; Baumert, Bos, & Lehmann, 2000), the three-track system was subject to a major debate. The findings revealed a particular overlap in student achievement between the three tracks and suggested a particularly strong correlation between students' achievement and socio-economic status (Neumann, Fischer, & Kauertz, 2010). After the first Programme for International Student Assessement (PISA) studies confirmed the results, the Länder individually started various efforts to reform their school system. This included attempts to introduce a longer period of comprehensive schooling by extending the Grundschule to 6 years and combining the Hauptschule and Realschule into one school form (with varying names). Today, a considerable number of the Länder have already implemented or intend to implement this type of two-track system. Here, after four years of comprehensive schooling in the Grundschule, students either start attending a *Gemeinschaftschule*, which essentially replaces the Haupt- and Realschule, or to the Gymnasium, if they intend to pursue an academic career.

HISTORICAL DEVELOPMENT OF THE GERMAN TEACHER PREPARATION SYSTEM

The historical development of the teaching profession and a formal system of teacher preparation is closely aligned with the development of the formal school system (Graf & Vogelsbacher, 2001; Sandfuchs, 2004, p. 15; Terhart, 2004). Teachers at early scholarly schools were mostly monks or priests (Sandfuchs, 2004), who had visited a scholarly school themselves and studied for some time (e.g., at a university; Graf & Vogelsbacher, 2001). Elementary school teachers, on the contrary, were mostly laymen: for example, cobblers, innkeepers, or other kinds of businessmen (Sandfuchs, 2004).

A first milestone in the development of the teaching profession and development of a formal teacher preparation system was the introduction of a state examination (examen pro facultate docendi) for those intending to teach at scholarly schools in two major German territories, Bavaria and Prussia in 1809 and 1810, respectively (Sandfuchs, 2004). According to Sandfuchs (2004) the examination represented an entrance examination, supposed to ensure a sufficient qualification of the teachers. In the beginning, the examination focused on the scholarly nature of the teacher as a whole; only later in the 19th century did state examination regulations require examinations in specific areas, such as philosophy, history, mathematics, science, religion, Hebrew, and newer languages (p. 18). In 1826, Prussia also introduced a probation year for teachers (Graf & Vogelsbacher, 2001), which was complemented by a preceding year of seminal learning at a school in 1890 (Sandfuchs, 2004).

Teachers at the elementary schools (including the early Volks- and Realschule) were not required to have any kind of specific preparation until late in the 19th century. These teachers often even only had limited skills in reading or writing. In light of these issues, the establishment of teaching seminars for elementary school teachers was enforced. From 1872 onward, the equipment used in these seminars and the education of the seminar teachers themselves improved considerably through substantial financial investments by the state governments. The situation for teachers at what was now essentially either a Volks- or Realschule, improved accordingly, both with respect to working conditions and payment. A refinement of the teaching seminars and, thus, the qualification (and situation) of teachers at the Volksschule continued until 1918. In the 1920s, the discussion about further improvement of the preparation of Volks- and Realschule teachers led to the introduction of higher education institutions, such as *Pädagogische Hochschulen* (universities of education), *Pägagogische Akademien* (educational academies), or the integration of teacher preparation into the regular universities in several territories. However, the idea did not find widespread acceptance across the territories and was reverted after 1933 (Sandfuchs, 2004, p. 20–24).

With the foundation of the Federal Republic of Germany in 1949, the Länder received (amongst others) authority over questions of educational politics: including the preparation of teachers (Sandfuchs, 2004; Viebahn, 2003). As a result, different teacher preparation systems developed.

Regular coordinatory meetings of the Secretaries of Education from the individual Länder ensured that the systems did not develop into systems too distinct from one another (Terhart, 2004). The beginnings were massively influenced by the post-Second World War situation, which was characterized by a shortage of qualified teachers and teacher educators. Teacher preparation had to be re-installed. In order to do so, some Länder re-introduced universities of education or academies of education set up in the tradition of the teaching seminars. This time was also characterized by an intense discussion on how teacher education should be organized. In 1969, the *Deutscher Bildungsrat* (German Education Council) proposed a threefold preparation program that included (a) training in educational and social sciences, (b) education in a scientific discipline (Fachwissenschaft) and aspects of education in this discipline (Fachdidaktik), and (c) practical experience. This proposal also included the idea of integrating teacher education into universities. In the years that followed, this integration was implemented in all but one of the Länder (i.e., Baden-Württemberg; Sandfuchs, 2004, pp. 27–29).

The reform process led to teacher education programs for Grundschule, Haupt- and Realschule, and Gymnasium offered at universities (or in the case of Baden-Württemberg for all school types except the Gymnasium at the Pädagogische Hochschulen). This structure mostly represents the current state of the teacher education system in Germany as described by Terhart (2004). Each of these programs consists of training in educational sciences, two subjects (or three in the case of Grundschule teachers), the respective subject educations (Fachdidaktiken), and practical studies in schools. Sometimes, two or more of these programs are combined into one, such as a teacher education program for Grund-, Haupt-, and Realschule teachers or Grund- and Hauptschule teachers (p. 39–40). The existing differences in length of the study programs have been adjusted to 3 + 2 years study programs in most of the Länder in the last few years. Still, the programs for Gymnasium teachers have a larger amount of subject courses, while the Grundschul-, Haupt-, and Realschul-/Gemeinschaftsschul-programs emphasize the pedagogy and the subject education to a higher degree.

Independent of school level or track, teacher preparation consists of two phases (Blömeke, 2006; Viebahn, 2003). The first phase—the university teacher preparation program—ends with a certificate, which qualifies students for the second phase (Viebahn, 2003). The second phase, the *Referendariat,* is an in-service training phase, in which students teach a reduced number of lessons at school, while simultaneously attending a *Studienseminar* (study seminar) run by the state (Terhart, 2004; Viebahn, 2003). At the end of the second phase, the students have to pass another examination, which formally qualifies them as teachers and entitles them to teach at the respective schools for which they were trained (Terhart, 2004; Blömeke, 2006).

The most recent and still ongoing reform of teacher preparation in Germany began in 1999 with the Bologna declaration (Blömeke, 2006). With the decision to unify university degrees across Europe in 2009, the Secretaries of Education of the European member states decided to adopt the consecutive bachelor, master, and doctorate system. Also, the European Credit Transfer and accumulation System (ECTS) was introduced, in order to allow an easier exchange of students between universities. The objective behind these measures was to enhance the competitiveness of the European higher education (Blömeke, 2006).

THE CURRENT TEACHER PREPARATION SYSTEM IN GERMANY

With the decision to unify European university degrees and to adopt a bachelor, Master, and doctorate system, teacher preparation in Germany also had to be re-organized. This posed and still poses a major problem, since, as described above, teacher preparation in Germany requires students to study two subjects and the consecutive bachelor master system was not designed for this. While historically being organized as one phase (despite an in-between examination after approximately four semesters), teacher preparation now has to be organized in two phases, with each phase leading to a certificate (bachelor and master). In particular, the requirement of polyvalence—that is the possibility for students from a given bachelor program to, in principle, be eligible to apply for any (related) master program—presents a challenge for the re-organization of teacher preparation. Since, for obvious reasons, students cannot be enrolled in two parallel programs at once, specific bachelor and master programs for teachers were created.

The required accreditation for teacher preparation still is the *Allgemeine Hochschulreife*, which can be obtained at Gymnasium schools after a total of 12 or 13 years of schooling (Blömeke, 2006). Depending on the university, the specific bachelor program and the Länder, matriculation typically leads either to a Bachelor of Arts or a Bachelor of Science (for the teaching profession). Certification requires the completion of the program. This entitles students to enroll for a respective Master of Education program and, in order to ensure polyvalence, requires the completion of additional classes for enrollment in regular Master of Arts or Master of Science programs. Typically, however, students enroll for a Master of Education program. Completion of this program is accredited as a master degree, although in some Länder (e.g., Bavaria) an additional examination is required.

After successfully completing their university studies, the students are eligible to apply for the second phase of teacher preparation, the in-service training phase. The length of this in-service training lasts from 18 months to 2 years. During this phase, the students regularly teach classes with an increasing teaching load. In their teaching, they typically receive feedback from mentors that are regular teachers at the same school. The students are also regularly visited by teacher educators observing the lessons, giving feedback and also grading the students' work. At the end of this phase of in-service teaching, teacher candidates present one lesson in each subject as an examination lesson. This assessment, together with an oral examination and a thesis, presents the Second State Examination. After obtaining the Second State Exam, students become fully qualified teachers and can apply to ministries and schools for an appointment. An overview of the current structure of the teacher preparation system in Germany is shown in Figure 2.1.

TEACHER EDUCATION PROGRAM STRUCTURE

Based on the structure of the teacher preparation system shown in Figure 2.1, universities offer specific teacher education programs. Following the tradition of German teacher education, each program includes education in two subjects and the respective subject educations. In addition to being prepared in two subjects, students also take classes in educational

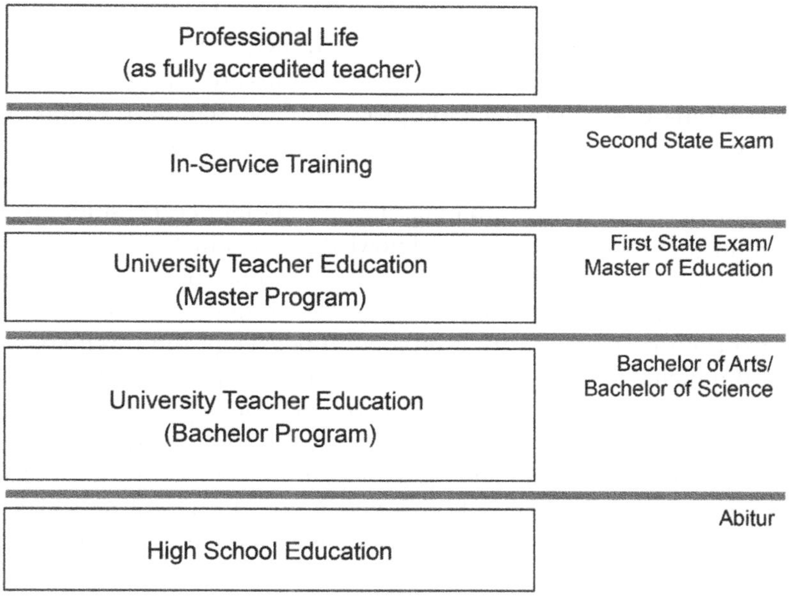

Figure 2.1 Stages and respective exams of the current teacher preparation in Germany.

sciences. These classes (can) comprise a variety of disciplines from educational science, including learning science to social science, depending on the teacher education program and, thus, the university. The general idea behind this threefold structure of teacher preparation programs is that the professional knowledge students need in order to be successful teachers includes content knowledge (CK), pedagogical content knowledge (PCK), and pedagogical knowledge (PK) (Baumert & Kunter, 2013; cf. Shulman, 1986, 1987). Students are expected to refine and deepen this knowledge in general school internships.

Despite the common features of teacher education programs outlined in the preceding paragraph, there are still considerable differences in the structure and content of teacher education programs across the Länder and even across universities within one Land. These differences include the distribution of ECTS points across classes in the subjects, the subject education and educational sciences and the length of school internships or the certification. For example, some universities exclusively offer a Bachelor of Arts after completing the two-subject bachelor program, while at others students receive a Bachelor of Arts or Bachelor of Science, depending on the combination of subjects.

In order to ensure a minimum comparability of teacher education programs across the Länder, these are based on common standards for teacher education agreed on by the Standing Conference of the Ministers of Education and Cultural Affairs (Sekretariat der Ständigen Konferenz der Kultusminister der Bundesrepublik Deutschland [KMK], 2004, 2008). These standards detail the professional competence students are supposed to develop during standard teacher education programs. The standards differentiate between the goals for bachelor and master degrees on the one hand and the in-service teaching period on the other hand. For educational sciences, the standards require that with obtaining the master degree, students have profound theoretical pedagogical knowledge. This includes, besides others, theories on teaching and learning, assessment approaches, different teaching methods or knowledge about socio-cultural influences on learning. Most of the standards are grounded in pedagogy and psychology, but some also include aspects of pedagogical content knowledge that students should be able to apply, using subject specific teaching approaches. They should, further, be aware of pre-knowledge, interest and motivation of students and how to integrate these into the planning of lessons (KMK, 2004).

These standards are the only common policy document on which the regulations for teacher education programs in the Länder are based. For the universities, in turn, the regulations together with the standards are the main resource based upon which teacher education programs are developed. The level of detail of these regulations varies across the Länder. At a minimum, it is specified for what type of teachers education programs are needed and how long the programs should be for each type of teacher. Other things addressed in the regulations include, for example, the length of general school internships or the integration of "heterogeneity and inclusion" (United Nations, 2006; Stöger & Ziegler, 2013; KMK, 2011).

In order to ensure compliance of the teacher education programs with the agreements of Bologna (i.e., European laws), the common standards of the KMK (2004, 2008) and regulations of the respective Land, the programs need to be accredited. In Germany, 10 different institutes are eligible to accredit new programs or re-accredit existing ones. These 10 institutes are, in turn, accredited by an accreditation council. The 10 institutes differ in their legal form: Some are foundations, some are associations and others are private organizations. There are two different ways of accreditation: a program and a system accreditation. For the program accreditation, the university prepares all documents that give information about the intent, structure, content, and assessments of specific studies. At the beginning, the accreditation committee will prove whether all legal aspects are correctly represented. Furthermore, the structure, content, and assessments are also evaluated. The committee will then certify the accreditation with or without obligations or even not certify it. For the system accreditation, the university prepares a concept on how to develop and ensure quality management structures for an accreditation within the university. This includes not only a description of how to ensure the quality of the internal procedures, but also the overall description of a specific program (e.g., the teacher education).

During the first decade of accreditation, one of the main aims was integrating the old structures into the new ECTS system. As in other countries, the ECTS points should be a measure of how much work is needed for a specific course. Overall, the bachelor degree corresponds to 180 points in the ECTS and the master degree to an additional 120 ECTS points. A typical teacher will receive his or her master degree after 5 years, organized in 10 semesters. A typical semester should, thus, include about 30 ECTS points, which is equivalent to 750–900 hours of work. This time span includes classes at universities as well as preparation of the classes or follow-up work. For the teacher education programs, the question of how the ECTS points should be divided up between the subjects, subject education, and educational science was intensely discussed. Especially the science and science teacher education programs require extensive work from the students beyond class attendance. The limitations that come with the introduction of the ECTS and total work hours thus lead to conflicts with respect to the time spent on subjects, subject education, and educational sciences. As a result, essentially every university found a different solution to the conflict leading to a huge heterogeneity of ECTS points or work hours, respectively, allocated to the subjects, subject education, or educational sciences as well as practical phases. Even the ECTS points for the bachelor and the master degree differ significantly between universities. This can be exemplified: The range of ECTS points needed to complete the bachelor and the master together lies between 75–120 ECTS points for each subject (with a mean of 100, found at nearly 50 universities), between 18 and 64 ECTS points in pedagogy, between 14 and 41 ECTS points for school internships and, finally, the ECTS points for the bachelor and master theses range between 20 and 40 (Centrum für Hochschulentwicklung, 2013 after DPG, 2014). A comparison across the Länder and universities shows that some universities focus on education in the subject, others on the subject education and educational science part. The majority of universities has implemented a somewhat balanced distribution of ECTS points across these fields.

SCIENCE TEACHER PREPARATION IN GERMANY

The preparation of (secondary level) science teachers is divided into the individual disciplines of science: biology, chemistry, and physics. Like in every other teacher preparation in Germany, science teacher preparation requires students to choose at least two subjects. Depending on the teacher education program, students may either freely choose their subjects or may be bound to a particular subject combination. That is, students can choose from biology, chemistry, and physics and either freely combine this subject with any other subject that is taught at school, or students might have to choose from a list of possible combinations (e.g., biology and chemistry, biology and physics, or mathematics and physics). Even when given the possibility to freely choose their second subject, however, students will usually choose a particular combination (e.g., mathematics and physics). This is partly due to the fact that universities will only guarantee for a very limited set of subject combinations to not pose any conflicts in the students' schedule(s) throughout the program.

For each subject, students take classes in the subject itself (i.e., biology) and classes on how to teach the subject (i.e., biology education). Teacher preparation programs typically start with introductory classes in the subjects (i.e., introductory physics classes), complemented by general education classes and classes on how to teach the subject. In order to achieve the required polyvalence in the bachelor program, the focus lies on classes in the subject with a few education classes and classes on how to teach the subject at most universities. Often, students are also required to do internships at schools. In the bachelor program at Kiel University, for example, (which requires a total of 180 ECTS points), students need to achieve 70 ECTS points in each subject (totaling 140 ECTS points) and 30 ECTS points in education related classes (plus 10 ECTS points for the bachelor's thesis). Hence, less than 20% of the total ECTS points in the bachelor program at Kiel University relate to educational sciences or education in subject education (see Figure 2.2). Of the 30 ECTS points required in education related classes, 10 ECTS points need to come from general education classes, 10 ECTS points from school internships, and 5 ECTS points relate to education in subject education.

The weighting of elements is reversed in the Master program, where the focus mostly lies on education/subject education (see Figure 2.3). In the Master program, up to 46 ECTS points are assigned to content specific classes (up to 23 ECTS points in each subject) and 56 ECTS points go to education-related classes (at least 10 ECTS points for education on subject education in each subject), including again 9 ECTS points for a school internship. Together with the 18 ECTS points related to the master's thesis, this leads to the required total of 120 ECTS points (Figure 2.3).

Teacher Preparation in Biology

Teacher preparation in biology is structured according to the aim of developing content knowledge (CK) and pedagogical content knowledge (PCK).[1] The central aim of teacher preparation in biology is to convey the basic qualifications needed to create successful teaching and learning

Two-Subject-Bachelor (BSc)		
Subject 1 (e.g. physics)		
Subject related courses	S 1-6	70 ECTS P
Subject 2 (e.g. mathematics)		
Subject related courses	S 1-6	70 ECTS P
Profile teacher education		
Education related courses	S 1-5	30 ECTS P
Bachelor thesis		
Subject 1 or 2 (duration 2 months)	S 5-6	10 ECTS P
Total Credit		180 ECTS P

Figure 2.2 Bachelor of Arts program at Kiel University. *Note:* S = semester, ECTS P = European Credit Transfer and Accumulation System Points.

Two-Subject-Master (MEd)		
Subject 1 (e.g. physics)		
Subject related courses	S 1-3	23 ECTS P
Subject education related courses	S 1-3	10 ECTS P
Subject 2 (e.g. mathematics)		
Subject related courses	S 1-3	23 ECTS P
Subject education related courses	S 1-3	10 ECTS P
Profile teacher education		
Education related courses	S 1-3	36 ECTS P
Bachelor thesis		
Subject 1 or 2 (duration 2 months)	S 4	18 ECTS P
Total Credit		120 ECTS P

Figure 2.3 Master of Education program at Kiel University. *Note:* S = semester, ECTS P = European Credit Transfer and Accumulation System Points.

processes in biology to the future teachers. In detail, this comprises the following knowledge and competences: The students

- display a substantial and expandable biological knowledge, analytical-critical thinking abilities, as well as methodological competencies;
- are familiar with the epistemology of the natural sciences and are able to apply this knowledge to biological contexts;
- are skilled in hypotheses-guided experimentation as well as hypotheses-guided comparison and systemizing;
- are able to handle school-relevant equipment to carry out biological research and experiments;
- are capable of comprehending biological circumstances in different contexts and can evaluate these by taking into account factual as well as normative arguments, while justifying the individual as well as societal relevance of biological issues;
- are able to design educational concepts and media for biology teaching and learning processes, to evaluate these and integrate current research findings from life sciences according to the students' pre-knowledge;
- feature an expandable PCK of biology, especially the fundamental findings of biology education research, of different curricular ap-

proaches, of students' (mis-)conceptions and barriers of learning in the different biological fields: also, they know the basics of designing and implementing biology lessons according to the national standards for biology teaching and learning;
- have the competences at their disposal to reflect, communicate, diagnose, and evaluate biological issues and are familiar with fundamental skills and scientific methods of biology education research; and
- possess an initial experience in designing and conducting biology lessons and know the basics on how to diagnose and evaluate students in the biology classroom.[2]

All future biology teachers have to develop CK in the traditional fields of biology (i.e., cell biology, morphology and physiology (plants and animals), neurobiology, the biology of behavior, genetics, molecular and developmental biology, evolution and systematics, and ecology as well as human biology). For students aiming at teaching at the Gymnasium, these areas are complemented by microbiology as well as immunology. Also all the other fields have to be studied in more depth than when studying for any of the other school types. The teaching formats for all of these contents are mainly lectures and practical work. Partly, students also attend seminars where they usually present current research papers from international journals.

Besides these traditional biological contents, biology teacher students are educated in fields of application in biology (e.g., health education, drug prevention [including psychological and physical aspects of drug abuse], and biotechnology as well as gene technology). Students also have to learn basic chemistry and physics that concern structures and functions of biological systems. In biology education, which predominantly takes place in seminars, the students get to know the fundamentals of the teaching and learning biology. Additionally, they have to cope with the conception and design of biology lessons, with the evaluation of students as well as their own teaching and they get an insight into biology education research.

Teacher Preparation in Chemistry

The foundation for the design of chemistry teacher education programs are the mentioned teacher education standards published by the KMK (2004), but also statement papers of Germany-wide organizations, such as the German Chemical Society. The KMK demands outcomes according to the following standards (translated from KMK, 2008): The future teachers have developed a solid knowledge base in chemistry and chemistry education that enables them to design teaching and learning processes in chemistry and to include new subject related and interdisciplinary developments in school curricula and school programs. They should be able to

- understand current chemical research;
- provide knowledge about topics and activities in chemistry related research institutes and industries;
- understand and evaluate chemical content in different areas of applications and represent them adequately;
- structure areas of chemical knowledge and applications according to relevant questions and link them to the topics of school chemistry;
- provide knowledge about important methods of scientific development and to carry out experiments safely;
- provide knowledge about historical developments of exemplary theories and concepts and their meanings;
- provide knowledge about the process of developing chemical knowledge and the relevance of chemistry for individuals and society;
- develop and evaluate conceptual approaches and suitable media for the teaching and learning of chemistry and adequately include new developments into their teaching;
- highlight and justify the importance of sustainable development for chemistry;
- provide profound knowledge in chemistry education as a foundation for future learning, especially about results of empirical research, conceptual and curricular approaches in chemistry education, as well as for diagnosing learning difficulties and pre-concepts with regard to competence-oriented teaching and learning in chemistry;
- provide knowledge about the design of learning environments with a special focus on heterogeneous pre-conditions and inclusive settings incorporating safety regulations for experimental environments; and
- provide first reflected experiences in planning and designing competence-oriented teaching of chemistry as well as the foundation of assessment and grading in chemistry classes.

Chemistry teacher education programs are usually organized according to the main areas of chemistry. These mainly include inorganic, organic, and physical/theoretical chemistry with additional fields of application depending on the university foci, such as polymer chemistry, biochemistry, and environmental chemistry. All areas include lectures, tutorials, and lab programs, the latter including safety courses. Excursions to chemical industry plants are also offered as talk series given by external guests, usually supported by the German Chemical Society.

The teacher education programs additionally include chemical education modules. The curriculum for these is less coherent across the universities. Rather, the standards depend on the emphases set forth by the chemical education faculty, which include professors and former teachers in most of the Länder. However, most students will get an introduction into the main elements of the teaching and learning of chemistry as a foundation. They participate in lectures and seminars on, for example, chemistry curricula and standards for chemistry education, research on students' preconditions such as pre-concepts, motivation, and interest, typical methods and media of chemistry classes, most important experiments and models, and the relevance of chemistry for different areas of the society.

Also lab courses focusing on experiments that can and should be carried out in school either by the teachers (demos) or by the students (hands-on inquiry) are very common. Those modules contain the handling (including safety aspects) and also the educational reflection of learning goals and challenges as well the curriculum implementation.

Another common curriculum area is the analysis, comparison, and use of different conceptual approaches linked to different curriculum emphases. These include inquiry-based approaches, historical or nature of science approaches, as well as context-based and socio-scientific issues approaches. The students learn to design classes according to the different approaches and to evaluate expected outcomes based on experience and research. The emphases on research in chemistry education completely depend on the researchers' fields of expertise. While some students might get an introductory insight into the field of empirical research in chemistry education, others might rather focus on design research and again others on the combination of both. The curriculum for chemistry teacher education at Kiel University, for example, is shown in Figure 2.4.

Teacher Preparation in Physics

Teacher preparation in physics includes classes in physics and physics education. The main aim is to develop a profound PCK on the basis of a sound CK and pedagogical knowledge (PK). The Standards for Teacher Education in Physics of the KMK (2008) envision that with obtaining their master or First State Examination, respectively, students

- have a profound content knowledge in physics that allows them to shape and assess physics-related instructional concepts and media, to follow recent developments in physics and integrate them into their instruction;

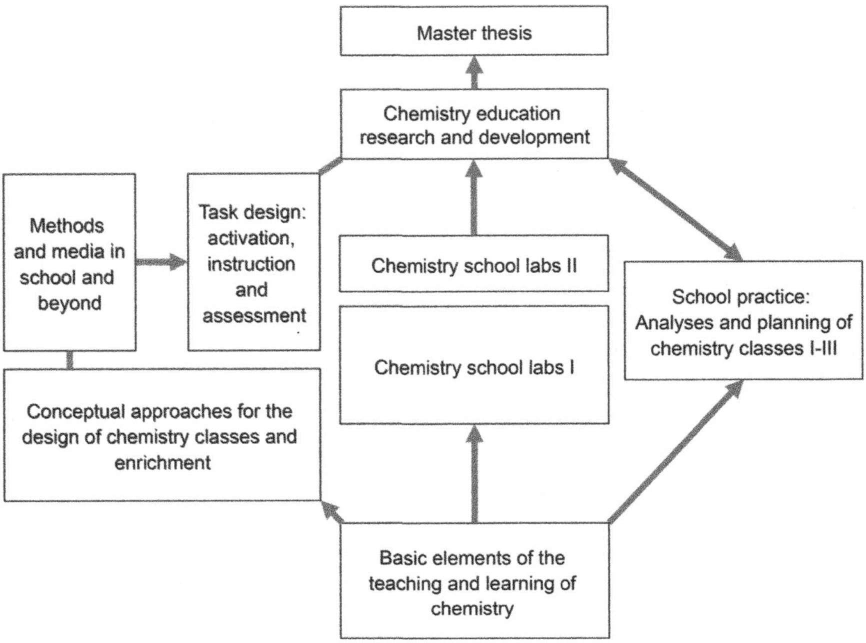

Figure 2.4 The chemistry teacher education program at Kiel University.

- are familiar with the working methods and procedures of physics and have the knowledge and skills needed to plan, perform and analyze experiments and to handle (school typical) experimentation devices;
- know the history of selected physical theories and terms as well as the process of knowledge acquisition in physics (knowledge about physics) and are able to define the societal relevance of physics;
- have a profound PCK, including the knowledge of physics education concepts, recent results of physics education research, typical learning difficulties or student conceptions in physics as well as possibilities to motivate students for learning physics;
- know possibilities for designing learning arrangements with respect to heterogeneous learning conditions and know the state of physics education research and development with respect to inclusive learning; and
- have first experiences in the planning and designing of instructional units as well as teaching individual lessons.

Moreover, the standards provide a list of contents to be covered in physics teacher education programs (depending on whether the program is directed at Haupt- and Realschule teachers or Gymnasium teachers). In experimental physics, these contents include mechanics, thermodynamics, electricity, and optics as well as atomic and quantum physics. Students studying to become teachers at the Haupt- and Realschule are, moreover, expected to obtain basic knowledge in solid state, nuclear and particle physics, as well as cosmology. Students studying to become teachers at the Gymnasium are expected to develop a more profound knowledge in these areas as well. In theoretical physics, students for the Haupt- and Realschule are only expected to obtain an overview of the structure and main concepts of physics, whereas teachers for the Gymnasium are expected to take classes covering theoretical mechanics, thermodynamics, electrodynamics, and quantum theory. In addition, students are expected to complete the basic labwork course in experimental physics and in school-oriented experimentation. Students studying for the Gymasium are expected to take an advanced labwork course in experimental physics. In applied physics, all students are required to obtain an overview of topics relevant for school, amongst them physics and sport, climate and weather, control and process technology, and the physics of sensory perception. Students aiming at teaching at the Gymnasium are expected to have a deep knowledge in at least one of these fields. In addition to physics, all students are expected to obtain some basic mathematical knowledge about vectors and matrices, functions, differential and integral calculus, differential functions, and statistics. Students wanting to teach at the Gymnasium are also expected to know about vector analysis, partial differential functions, Hilbert spaces and non-linear dynamics. Finally, in physics education, all students are expected to learn about physics education theories and conceptions, students' motivation and interest, learning processes and diagnosing learning difficulties, planning and analyzing instruction, physics tasks, experiments and other media, how to handle heterogeneity in physics instruction, and recent physics education research.

Accordingly, classes in physics typically include lessons on experimental physics (mechanics and relativity theory, optics, electricity, thermodynamics, nuclear and quantum physics, and solid state physics), theoretical physics (mechanics, electrodynamics, thermodynamics, and quantum physics), and lab work classes. Classes in physics education are fairly diverse across universities. They, however, typically include content on the fundamentals of physics education, including topics such as the goals of physics education (i.e., standards and other policy documents), student conceptions, and the nature of science.

In their 2014 report, the German Physics Association attempted to synthesize the main goals and contents of typical physics teacher education programs into a list of what students should have learned, acquired, or should be capable of by the end of the master program. According to this report, a student should

- have acquired a solid physical knowledge beyond the knowledge obtained from school and is familiar with the discipline's structure including theoretical physics and the historical development as well as the Nature of Science;
- be able to classify physical effects and describe them using mathematical models;
- make use of his or her knowledge in analyzing physical effects from everyday phenomena and implement this in the classroom;
- be able to design, build, and perform experiments for the classroom and interpret results correctly;
- have understood the idea of overarching concepts, such as energy, and take these into account in the instructional design;
- be familiar with the applications of physics in engineering and scientific research and can use this knowledge;
- take students' pre-knowledge, typical learning difficulties, motivation, and interest into account when planning teaching with respect to individual needs;
- have knowledge on didactic and educational psychology research of learning in physics, additionally taking gifted students into account;
- be able to select content and goals due to the situation and objectives of physics teaching;
- know different methods of diagnosis and feedback within learning progress;
- make decisions while teaching based on didactic theories and structuring approaches;
- be able to plan and design learning environments for self-directed learning;
- be capable of analyzing his or her own and others' teaching and reflect it critically;
- be able to further study technical and didactic knowledge and deepen his or her understanding of it and be familiar with relevant learning strategies (lifelong learning); and
- have key skills such as learning and working techniques, communication skills, ability to access literature, etc.

Nevertheless, how students are taught this knowledge and these skills varies considerably across teacher education programs (more so in physics education than in physics). The teacher education program at Kiel University, for example, includes (the typical) classes in mechanics and thermodynamics, electricity, magnetism and optics, atom and quantum physics, nuclear, particle, and astrophysics as well as theoretical mechanics and electrodynamics. Students also take the basic and advanced level lab work course and classes in mathematics (for physicists). In addition, students need to take classes in molecular and surface physics, computer sciences, an integrated course on physics, energy and environment and, prior to taking the basic level lab work course, an electronics and measurement lab work course. In physics education, all classes are thought to build an empirically grounded understanding of how scholars learn physics. Students training to become teachers typically take a class on the foundation of physics education (covering topics, such as the goals of physics education, models of physics learning, motivation and interest, or typical methods, such as cooperative learning and media, e.g., the experiment). In addition, students need to take a class on the planning, implementation, and analysis of physics lessons. Further courses require reading physics education research papers or taking specific lab work courses.

SUMMARY AND DISCUSSION

The German school and teacher preparation system is influenced by the fact that Germany is a federal state. This led to the development of sixteen somewhat different set-ups. These systems, however, share what can be considered a common core. In all teacher preparation systems, the Allgemeine Hochschulreife (i.e., general qualification for university entrance) is the requirement for entering a teacher preparation course (for a description of the school systems see for example Döbert, 2007). Also, teacher preparation is organized in two phases in all systems. The first phase consists of education at the university, the second phase of an in-service training. Successful completion of the university teacher education program is the main requirement for entering the in-service training. Historically, teacher education programs at the university as well as the in-service training phases differed slightly in formal organization and how the contents were taught. The biggest difference was to be found in the teacher education, which was organized at the so-called Pägogische Hochschulen in some Länder and that teacher education programs for becoming a teacher at Haupt- or Realschule had a stronger focus on subject education and educational science classes than teacher education programs for those wanting to become teachers at the Gymnasium.

With the mediocre performance of German students in the Trends in International Mathematics and Science Study (TIMSS) and the Program for International Student Assessment (PISA), many of the Länder initiated reform efforts of their school systems, which also manifested themselves in teacher and science teacher education (for details see for example

Neumann et al., 2010). One of these reforms included the reduction of a three-track school system in middle school with only one track leading to an academic career to a two-track system in which both tracks can lead to academic training. The biggest impact on the German teacher preparation system, however, was the Bologna declaration (European Ministers of Education, 1999). Following the declaration, decisions were needed to be made regarding how a teacher preparation system focusing on the education in two subjects, the respective subject education and educational sciences, can be transformed into a consecutive system of undergraduate (bachelor) and graduate (master) studies. This process was not centrally organized, neither on the federal level nor on the level of the individual Länder. In fact, many universities developed their own solutions, which led to a tremendous increase in variety amongst teacher education programs, rendering one of the central ideas of the Bologna reform, the idea of polyvalence, void. Prior to Bologna, it was possible to change from one university to another. Sometimes, students needed to take additional classes, but most of the time classes they had taken at their previous university where accredited. After Bologna, classes became increasingly different and located at different stages of teacher education programs. Thus, changing from one university to another before completing the program nowadays is, de facto, impossible.

Although the KMK (2004, 2008) standards for teacher education provide a common basis for the Länder's regulations, teacher education programs are not necessarily aligned accordingly. This is mostly due to the fact that the standards focus on the end of the master program, while the exact time when topics are taught can vary from program to program. However, despite the standards listing a set of contents to be covered, there are also differences in the extent to which these contents are covered. For example, the standards require students to obtain a broad overview of topics, such as physics and sport, climate and weather, or physics of sensory perception. At most universities, there are no dedicated classes covering these topics; they will be addressed in regular physics classes (i.e., sport examples will be discussed in the mechanics class), in applied physics classes, or sometimes in integrated classes such as the class on physics, energy, and environment at Kiel University. As a result, the amount of teaching students receive on these topics and, thus, the extent to which students meet the expected standards may vary considerably. So far, however, there is no debate on this variation, partly because there is no assessment of the standards, neither in the accreditation process (i.e., through curriculum analysis by the accreditors), nor through Länder-wide or national tests. This, in turn, is most likely due to a lack of high quality instruments for assessing the standards.

The second phase of teacher education has also changed with time, but mostly with respect to formal features, such as the length of studies and the teaching load. This is somewhat surprising, as the division of teacher education into an academic and a practical stage of the courses of studies and the conduct at two different institutions (university school, "study seminar") has not always been agreed upon unanimously. The Hochschulrektorenkonferenz (1998), for example, noted a need for regular exchange of personnel in order to avoid a separation of the two "cultures of education systems": Experienced teachers, particularly when participating in the practice training for post-graduate students, should receive further academic education so that they can establish closer relations between theory and practice. They further suggested that professors of the theory of teaching in a subject and of educational science should complete a praxis semester at an education-related research project in a school approximately once every four years. Both were suggestions that did not find widespread acceptance. Currently, this kind of exchange of knowledge is mostly realized by (a) offering teachers the possibility of working at a university in the subject (e.g., chemistry) education department for three to five years (depending on the Land) at full payment, and (b) some Länder or universities requiring university professors to have a minimum of three years of teaching experience at a school. Both solutions, however, do not lead to the envisioned continued exchange and prevention of the development of two cultures.

Another outcome of the debate around the excessively strong separation between theoretical and practical education was that many Länder required more and longer general school internships in the teacher education programs. However, up to now, little to no evidence has been presented that this in any way led to a more refined professional knowledge or competence of students at the end of the teacher education program. The available evidence, in fact, suggests that a deep integration of the general school internships into the program (i.e., a thorough preparation of students for the complex requirements of observing and teaching and professional support by teacher educators) is more beneficial than increasing the number or length of general school internships (Schubarth et al., 2011).

In the case of science education, a specific impact of recent developments on science education has come in the form of changes regarding the tracking system. With the introduction of essentially one non-academic (Gemeinschaftsschule) and one academic track (Gymnasium), a trend towards teaching science as a single subject at the non-academic track and retaining the three distinct subjects biology, chemistry, and physics at the academic track has developed. While this may seem reasonable at first sight, it creates two problems: First, new teacher education programs need to be established by teaching integrated science and science education, which in turn leads to an even greater diversification of teacher education programs. Second, there is a general concern that this may lead away from one of the strengths of the German teacher preparation system, which is a strong focus on the development of a profound content knowledge in the subjects.

In summary, the German science teacher preparation and, more specifically, the teacher preparation in biology, chemistry, and physics, is deeply rooted in respective disciplines. Students teach two subjects, which may both be subjects of natural sciences, but do not have to be and receive a sound education in both of them. In addition, students receive tuition in the education of both subjects (i.e., biology education) and in educational sciences. While there is a particular consensus that the education in the subjects is of high quality and teachers typically develop a profound CK, there is less agreement on the development of PK and PCK. This is mostly due to the fact that it is not clear up until now, how students develop PCK from CK and how both PK and PCK play out in teaching. The question of a better integration of general school internships and the first and second phase of teacher education remains unanswered and to be addressed in the future.

NOTES

1. Since teacher education programs (at the same university) typically require students to take the same classes in educational science or choose from the same set of classes, we will only focus on content and pedagocial content knowledge in the discipline-specific sections.
2. This aspect applies to the internships teacher students pass during their university studies. These are supervised by university lecturers and are accompanied by PCK seminars.

REFERENCES

Baumert, J., & Kunter, M. (2013). The COACTIV model of teachers' professional competence. In M. Kunter, J. Baumert, W. Blum, U. Klusmann, S. Krauss & M. Neubrand (Eds.), *Cognitive activation in the mathematics classroom and professional competence of teachers. Results from the COACTIV project* (pp. 25–48). New York, NY: Springer.

Baumert, J., Lehmann, R., Lehrke, M., Schmitz, B., Clausen, M., Hosenfeld, I., & Köller, O. (1997). *TIMSS. Mathematisch-naturwissenschaftlicher Unterricht im internationalen Vergleich* [TIMSS. An international comparison of athematics and science classes]. Deskriptive Befunde. Opladen, Germany: Leske+Budrich.

Baumert, J., Bos, W., & Lehmann, R. (2000). *TIMSS/III. Dritte Internationale Mathematik- und Naturwissenschaftsstudie: Mathematische und naturwissenschaftliche Bildung am Ende der Schullaufbahn* [TIMSS/III. The third International Mathematics and Science Study: Mathematics and science education towards the end of the school career]. Opladen, Germany: Leske+Budrich.

Blömeke, S. (2006). Globalization and educational reform in German teacher education. *International Journal of Educational Research, 45*(4–5), 315–324. Retrieved from http://doi.org/10.1016/j.ijer.2007.02.009

Centrum für Hochschulentwicklung. (2013). *Monitor Lehrerbildung*. Retrieved from www.monitor-lehrerbildung.de

Deutsche Physikalische Gesellschaft (DPG). (2014). *Zur fachlichen und fachdidaktischen Ausbildung für das Lehramt Physik* [On the content and pedagogical content education of physics teachers]. Bad Honnef, Germany: DPG.

European Ministers of Education(1999). *Joint declaration of the european ministers of education*. Bologna, 19 June 1999.

Graf, F., & Vogelbacher, B. (2001). Lehrer/Lehrerin [Teacher]. In L. Roth (Ed.), *Pädagogik. Handbuch für Studium und Praxis* [Pedagogy. Handbook for study and practice] (pp. 1141–1157). München, Germany: Ehrenwirth.

Hochschulrektorenkonferenz. (Eds.) (1998). *Empfehlungen zur Lehrerbildung* [Recommendations on Teacher Education]. Retrieved from http://www.hrk.de.

Neumann, K., Fischer, H. E., & Kauertz, A. (2010). From PISA to educational standards: The impact of large-scale assessments on science education in Germany. *International Journal of Science and Mathematics Education, 8*(3), 545–563.

Sandfuchs, U. (2004). Geschichte der Lerhrerbildung in Deutschland [History of teacher education in Germany]. In S. Blömeke, P. Reinhold, G. Tulodziecki, & J. Wildt (Eds.), *Handbuch Lehrerbildung* [Handbook of teacher education] (pp. 37–59). Bad Heilbrunn, Germany: Klinkhardt.

Schubarth, W., Speck, K., Seidel, A., Kamm, C., Kleinfeld, M., & Sarrar, L. (2011). Evidenzbasierte Professionalisierung der Praxisphasen in außeruniversitären Lernorten. Erste Ergebnisse des Forschungsprojektes ProPrax [Evidence based professionalization of practical phases in out-of-university learning environments. First results from the ProPax project]. In Schubarth, W./Speck, K./Seidel, A (Hrsg.). (2011). *Nach Bologna: Praktika im Studium—Pflicht oder Kür? Empirische Analysen und Empfehlungen für die Hochschulpraxis* [After Bologna: Internships during university studies—Obligation or option?]. Potsdam, Germany: Universitätsverlag, S. (pp. 79–212).

Sekretariat der Ständigen Konferenz der Kultusminister der Bundesrepublik Deutschland [KMK]. (2004). *Standards für die Lehrerbildung: Bildungswissenschaften (Beschluss der Kultusministerkonferenz vom 16.12.2004)* [Standards for teacher education: Educational sciences (resolution of the Conference of the Secretaries of Education) from 2004-12-16]. Retrieved from http://www.kmk.org/fileadmin/veroeffentlichungen_beschluesse/2004/2004_12_16-Standards-Lehrerbildung.pdf

Sekretariat der Ständigen Konferenz der Kultusminister der Bundesrepublik Deutschland [KMK]. (2008). *Ländergemeinsame inhaltliche Anforderungen für die Fachwissenschaften und Fachdidaktiken in der Lehrerbildung (Beschluss der Kultusministerkonferenz vom 16.10.2008)* [Common content requirements for the sciences and education in the sciences in teacher education for the Länder (Resolution of the Conference of the Secretaries of Education) from 2008-10-16]. Retrieved from http://www.kmk.org/fileadmin/veroeffentlichungen_beschluesse/2008/2008_10_16-Fachprofile-Lehrerbildung.pdf

Sekretariat der Ständigen Konferenz der Kultusminister der Bundesrepublik Deutschland [KMK]. (2011). *Inklusive Bildung von Kindern und Jugendlichen mit Behinderungen in Schulen (Beschluss der Kultusministerkonferenz vom 20.10.2011)* [Inclusive education of children and adolescents with special needs in schools (Resolution of the Conference of the Secretaries of Education) from 2011-10-20]. Retrieved from http://www.kmk.org/fileadmin/veroeffentlichungen_beschluesse/2011/2011_10_20-Inklusive-Bildung.pdf

Shulman, L. (1986). Those who understand: Knowledge growth in teaching. *Educational Researcher, 15*(2), 4–14.

Shulman, L. S. (1987). Knowledge and teaching of the new reform. *Harvard Educational Review, 57,* 1–22.

Stöger, H. & Ziegler, A. (2013). Heterogenität und Inklusion im Unterricht [Heterogeinity and inclusion in teaching], *Schulpädagogik Heute, 4*(7), 1–31.

Terhart, E. (2004). Struktur und Organisations der Lerhrerbildung in Deutschland [Structure and organisation of teacher education in Germany]. In S. Blömeke, P. Reinhold, G. Tulodziecki, & J. Wildt (Eds.), *Handbuch Lehrerbildung* [Handbook of teacher education] (pp. 37–59). Bad Heilbrunn, Germany: Klinkhardt.

United Nations (2006). *Note by the Secretary-General: Final report of the ad hoc committee on a comprehensive and integral international convention on the protection and promotion of the rights and dignity of persons with disabilities, 6 December 2006, A/61/611.* Retrieved from http://www.refworld.org/docid/45c30c560.html.

Viebahn, P. (2003). Teacher education in Germany. *European Journal of Teacher Education, 26*(1), 87–100.

CHAPTER 3

SECONDARY SCIENCE TEACHER EDUCATION IN FINLAND

Jouni Viiri
University of Jyväskylä

Tiina Silander
University of Jyväskylä

FINNISH EDUCATIONAL SYSTEM

Finnish society is founded upon a high level of education and expertise. A free and equally beneficial education, available to everyone regardless of gender, residential status, or socioeconomic background, is one of the cornerstones of the Finnish educational policy. Free education means that in preprimary and basic education all learning materials, daily meals, and transportation for pupils who live further away comes at no cost to families.

The Finnish system of basic education was reformed in the 1970s. With this Comprehensive School Reform (Simola, 2005), basic education and teacher education changed. The dual-track school system was replaced with 9 years of compulsory school in which all pupils take the same curriculum until the end of 9th grade. Thus, there is no streaming, and pupils study

Model Science Teacher Preparation Programs, pages 53–67

in heterogeneous groups in every subject, including science. The basic ideology in Finnish education is that the potential of each pupil should be maximized. Therefore, all pupils have the right to general support, which means high-quality education as well as guidance and support during all their studies. The purpose is to prevent existing problems from becoming serious. For example, if a pupil has been sick and thus not able to participate in teaching, he/she has a right to get remedial teaching to support his/her learning. Preprimary education is offered to all 6-year-olds. In preschool, pupils are taught basic skills, knowledge, and capabilities from different areas of learning. Learning through play is the most prominent feature of Finnish preschool education.

The educational period from the 1st to the 6th grade of compulsory school forms the lower school, where most of the daily school subjects are taught by a single person, the class teacher. Typically, this person teaches a specific group of pupils for several years in a row. The grades from the 7th to the 9th form the upper school, where different school subjects are taught by individual subject teachers (i.e., one teacher for languages such as English and Swedish, another for natural sciences such as physics and chemistry, etc.). After completing compulsory school, pupils choose to continue to either upper secondary school or to begin vocational education.

In addition, the schools follow the national core curriculum (Basic Education, n.d.), which is developed by the National Board of Education closely with other stakeholders such as municipalities, universities, and teachers. The national core curriculum is not a document merely handed down for teachers to implement, but teachers are involved in designing it. The curriculum for basic education specifies the objectives and core content of subjects, based on which, the municipalities and schools develop their own curricula. Thus, there is room for local or regional specificities. This is one example of the decentralized decision making and local responsibility that have been characteristics of Finnish education policy since the 1980s.

Municipalities, schools, and teachers have much freedom in terms of organizing education and teaching. Autonomy and trust are two distinctive features of Finnish education policy. Finnish teachers are trusted to be academic professionals who have good theoretical and practical education, high morals, and a strong ethical commitment to their work (Niemi, Toom, & Kallioniemi, 2012). Finnish teachers are able to organize their teaching autonomously in terms of teaching and learning methods or materials to be used. They also have a very active and important role in curriculum development. For instance, during the process of national curriculum reform teachers are asked to comment the draft versions of the curriculum and based on the national curriculum teachers develop their own school curriculum. Teachers are also responsible for pupil assessment since the main type of pupil assessment is the continuous assessment during the course of studies. Continuous assessment is to guide and help pupils in their learning process. Assessment is based on the objectives written in the national curriculum where examples of good knowledge are given (good knowledge means grade 8, in Finland grades are from 4 to 10). As an example in physics education one goal stated in the national curriculum is "to guide the student to use different models in the description and explanation of phenomena and in forecasting" (Finnish National Board of Education, 2014, p. 390). Related to this goal, the advice given to teachers for grade 8 is the following: "the student can use simple models and make forecasts as well as practices making simple models from measurement results. The student can describe a model and name its restrictions and limitations" (Finnish National Board of Education, 2014, p. 451).

There are no inspection systems or centrally organized nationwide tests in basic education. The National Board of Education organizes sample-based and thematic evaluations (e.g., science) to develop education and to provide information for policy decisions. At the end of upper secondary education, students take the national matriculation examination, which is the only standard national test in Finland.

Autonomy and pedagogical freedom rely on high-quality teacher education. Teacher education was transferred to universities in 1971. The purpose of this change was to unify elementary and secondary education into a single entity and to develop a high academic standard for prospective students. Before then, class teachers were educated at teacher colleges (seminaries), and subject teachers were trained at universities. The latest reform at the beginning of the millennium aimed at further reinforcing the academic basis of the teaching profession. The new curriculum for teacher education emphasizes teachers' readiness to apply research knowledge in their daily work. Research-based teacher education leans on the idea that teachers' professional development should be supported as a process that continues throughout their career (Silander & Välijärvi, 2013).

Today, the basic qualification for class teachers and subject teachers is a master's degree. Class teachers major in education, and faculties of education are responsible for organizing class teacher education programs. Subject teacher programs are organized cooperatively between the department of the teaching subject and faculties of education. Subject teachers major in the academic subject to be taught.

Science Teaching in Finland

During the lower school period of compulsory school (Grades 1–6), natural sciences are taught by a class teacher, whose pedagogical studies contain between 5 and 10 European Credit Transfer and accumulation

System (ECTS, a European standard) of studies of natural sciences courses, depending on the university. For a teacher student, 1 ECTS equals about 27 work hours. According to the national core curriculum, the school subjects of biology, chemistry, physics, geography, environmental studies, and health education form a combined study package called environmental studies. In secondary school, mathematics, physics, chemistry, biology, and geography are studied as separate subjects. Typically, science teachers specialize in teaching either mathematics, physics, and chemistry or biology and geography. Thus, they must have studied all subjects they will teach.

TEACHER STUDENT SELECTION AND TEACHER QUALIFICATIONS

The basic qualification for science teachers is a master's degree (300 ECTS). The teaching profession is highly valued, and thus, most teacher education programs receive far more applicants than there are study places available.

Since after completing the teacher education program teacher students are fully qualified teachers and have the same status as all other teachers, the student selection process plays an important role. It is very important to have already tested applicants for their suitability for the teaching profession before they are admitted to teacher education programs. Such a test is a mandatory prerequisite for admission, even when there is a shortage of applicants. The test screens out unsuitable candidates, and is comprised of interviews, where a student's aptitude for the job of a teacher is measured according to a variety of interactive skills including pedagogical interests and orientation toward a lifetime of learning. The interviews and subsequent evaluations that take place during these tests include faculty members from the teacher education department and subject-centric departments.

The old model for subject teacher education was that secondary school subject teachers had academic discipline education (about 5 years) in universities and thereafter practical teacher training in practice schools connected to universities. This model was rejected as it does not support integration of pedagogical and subject knowledge. In subject teacher programs today, there are two models. The more traditional model includes bachelor's and master's degrees in the teaching subjects. After completing these degrees, the student can apply for pedagogical studies. In integrated programs, the subject studies and pedagogical studies are integrated from the outset. In this model, the student applies to the subject teacher program when he or she enters the university.

The competition among natural sciences applicants differs depending on the specific major subject. For example, in 2014, the University of Jyväskylä admitted about 20% of the students who applied for physics, 15% of the students who applied for chemistry, and less than 5% of the students who applied for biology (Hakijamäärät, n.d.). Only a portion of the admitted students will in turn be chosen for the pedagogical study program. At the University of Jyväskylä, students can apply to this additional program either at the beginning or later during their studies.

After graduating from a university, students are licensed as teachers and may apply for teaching positions in schools. Education providers are responsible for employing their teaching staff. There is an open market for teacher positions, and vacant positions are shown on Internet sites, newspapers, and teacher union journals. No extra courses, tests, or diplomas are needed. Teachers do not need any recertification over time (e.g., after 5 or 10 years of service) or while moving from one area of the country to another.

One indication of the high level of teacher education is that teachers are able to continue for doctoral studies. The PhD thesis very often deals with subjects' pedagogy, for example, pedagogical content knowledge (PCK), developing design-based teaching sequences. At the University of Jyväskylä, every year about seven teachers complete PhD studies in the teacher education department. A doctorate degree is the principal degree in high-level university education.

After completing a master's degree, the teacher is fully qualified. The drawback is that then developing his or her expertise in pedagogy or in the subject is more or less his or her own responsibility. According to the official rules, teachers must participate in in-service training for 3 days per year. There is no general plan commanding what this education should involve, and there are no national, regional, or school restrictions. The supplemental training of teachers nationally is random and short-sighted, and the organizational structure requires full-scale development. The Ministry of Education and Culture has provided funding for the Osaava program, which aims at nationally developing the skills of educational personnel. The program's goals are to activate schools and the know-how of the faculty members who teach at different levels of education. The program provides local and regional support for developing the structures, plans, and educational approaches that require reforming. Finnish teachers are trusted in terms of their expertise in education and pedagogy. Their work is not subjected to evaluation. For instance, headmasters do not typically enter classrooms to observe lessons. The salary that teachers receive is not tied to any type of evaluation process, either. However, recently, Finnish teachers have seen an increasing trend of developmental conversations, which are supposed to support the professional development of teachers. Teacher quality is a result of careful quality control at entry into teaching instead of measuring teacher effectiveness in service. In addition, teacher education aims to educate teachers who are lifelong learners and motivated to improve in their profession.

GENERAL PRINCIPLE AND STRUCTURE OF TEACHER EDUCATION

The structure of teacher education programs, in general, is the same at all universities since the basic structure is regulated by law. Of course, there are some differences in course content since the universities are autonomous, and they can decide their own curricula details. Principles are based on recommendations by the Ministry of Education and Culture, by national working groups related to teacher education reforms such as the Bologna (The Bologna, n.d.), or an agreement between the deans of the faculties of education.

Finnish teacher education is committed to the development of a research-based professional culture. The basic aim is that teachers learn and internalize a research-oriented attitude toward their work. This means that teachers learn to take an analytical and open-minded approach to their work, that they draw conclusions based on their observations and experiences, and that they develop their teaching and learning environments in a systematic way.

Teachers need deep knowledge of the most recent advances in research in the subject they teach. In addition, they need to know the latest research concerning teaching and learning. The critical scientific literacy of teachers and their ability to use research methods are crucial. Thus, teacher education programs require study of qualitative and quantitative research traditions. The aim of these studies is to train students to analyze problems they may expect in their future work. Professors are responsible for guiding students into the research-oriented aspect of their education. The aim is to engage students to become active participants in the education society.

The goal of teacher education is to equip teachers with research-based knowledge and skills and methods for developing teaching, cooperating at school, and communicating with parents and other stakeholders. Thus, teacher education combines theoretical knowledge with practical know-how. In practice, this is achieved through the supervised teaching practice courses, which are an important part of standard pedagogical studies. The aim is to support students in their efforts to acquire professional skills in researching, developing, and evaluating teaching and learning processes. In addition, this is one place for students to reflect critically on their own practices and social skills in teaching and learning situations. The main principle is that practice should start as early as possible and support student teachers' growth toward expertise. At the beginning, students observe school life and pupils from an educational perspective, and later they focus on specific subject areas and pupils' learning processes. Finally, student teachers take holistic responsibility in their teaching as students plan, implement, and evaluate broader learning modules

Responsibility for subject teacher training lies partly with the faculties of education and partly with the faculties of different teaching subjects. Secondary-level subject teacher students are students enrolled within the department of their main subject. For instance, physics teacher students are members of the physics department. If a student has passed the entrance test, he or she can choose either the teacher education line or the physicist line. The subject departments are responsible for the students' content-centric studies while the Department of Teacher Education is responsible for the students' pedagogical studies. The teacher training school associated with the university will also take part in the arrangement of the pedagogical studies. The cooperative base of teacher education relies on this three-sided arrangement presented in Table 3.1.

CONTENT OF TEACHER EDUCATION

The profession of a teacher is regulated, and teacher qualifications are defined in legislation (Ministry of Education and Culture, n.d.). Main regulations for teacher education are stated in the Universities Act (#558/2009) and the Government Decree on University Degrees (#794/2004) which also contains as an appendix the Teachers' Education Decree which lists

TABLE 3.1 Main Components of Teacher Education Programs for Secondary School Teachers (Niemi & Jakku-Sihvonen, 2006)

Secondary School Teacher Education Program	Bachelor's Degree 180 ECTS	Master's Degree 120 ECTS	Total 300 ECTS
Academic studies in different disciplines (major)	60 (including a BA thesis 6–10)	60–90 (including an MA thesis 20–40)	120–150
Pedagogical studies (minor)	25–30 (including supervised teaching practice)	30–35 (including supervised teaching practice)	60
Academic studies in different disciplines (1–2 minors)	25–60	0–30	25–90
Language and communication studies, including ICT, optional studies	35–40	0–30	35–70

the faculties that provide teacher education. Besides, the Teaching Qualifications Decree (#986/1998) prescribes the qualification requirements for educational staff (Sähköinen säädöskokoelma, n.d.).

Subject teacher training includes studies in two to three teaching subjects, along with the teachers' pedagogical studies as part of the master's degree. A teaching subject means a subject included in the curriculum of the lower or upper secondary school, for example, physics or chemistry. Teacher students' studies in a teaching subject do not in general differ from the courses for other students. For instance, the students aiming for the physics teacher degree enroll in the same physics courses as the students studying for the physicist line. There are also some special courses for the physics teacher students, for example, physics school demonstrations, history of physics, and physics PCK courses.

In a typical situation, a student majors in physics, chemistry, or mathematics and minors in at least other two disciplines mentioned. This subject combination design is fueled by the practical situation that in general the same mathematics and science (physics and chemistry) teachers teach these subjects at schools. Therefore, students study all three subjects and have obtained more than the minimum 300 study points when they leave the university. The other typical combination is to major in biology or geography and minor in the other.

Science teacher students write their master's thesis at the faculty of science and learn research methodology for scientific operations. For instance, for physics teacher students, the thesis can deal with pure physics or pedagogy of physics. Science teacher education also includes research studies, communication and language studies and information and communication technology (ICT) studies.

Pedagogical Studies

All teacher education programs include pedagogical studies that (minimum of 60 ECTS) are obligatory for all teachers as a minor subject. The pedagogical studies are approximately the same for primary and secondary teachers as well as vocational and adult education teachers. These studies give a formal pedagogical qualification to teachers at all levels in the Finnish educational system regardless of the program in which they are enrolled. According to legislation, pedagogical studies must be studies in the science of education with an emphasis on didactics. Teaching practice (minimum of 20 ECTS) is part of pedagogical studies. In addition to academic subjects and pedagogical studies, teacher students can choose special education studies for his or her master's degree.

In addition to studies in the teaching subjects, students in the teacher line study pedagogy. These pedagogical studies are completed in the Department of Teacher Education at the faculty of education. The scope of the pedagogical studies is 60 ECTS (the subject study module in education), and students include them in their degree as a minor subject.

Since universities are autonomous, there can be slight differences in the content of pedagogical studies. The topics in the pedagogical studies at the University of Jyväskylä are described in Table 3.2. These studies provide students with the pedagogical capabilities required for independent performance of teaching duties at comprehensive schools, upper secondary schools, and other educational institutions. Strong theoretical thinking, reflection on and evaluation of one's own actions and development, as well as guidance toward good teaching practices are very important in the studies. The objective is for students to develop into teaching professionals who will develop their own work and working community.

The pedagogical studies are divided into the basic study courses completed within a bachelor's degree (25 ECTS) and the subject study courses completed within a master's degree (35 ECTS). The aims of the basic pedagogical studies are that upon completion of this study module, students will perceive pedagogical phenomena from different theoretical and practical perspectives, identify their experiences, conceptions, feelings, and working habits in interaction, are able to examine the construction of expertise in the educational field, and are able to analyze educational phenomena scientifically. Courses in basic pedagogical studies include five modules that are described in more detail in the following section.

TABLE 3.2 Pedagogical Courses in the Subject Teacher (Lower and Upper Secondary School) Studies at the University of Jyväskylä

	Bachelor			Master	
	Study Year 1	Study Year 2	Study Year 3	Study Year 4	Study Year 5
Scientific Thinking and Knowledge					
Interaction and Collaboration		15		5	
Education, Society and Change				4	
Learning and Guidance		5		5	
Constructing Scientific Knowledge: Teacher as Researcher				5	
Studies in Education Administration				1	
Training in the practice school		5		15	

Basic Pedagogical Studies

Interaction and Cooperation. Students learn to observe different interactional situations, analyze them from selected viewpoints, and examine the feelings aroused by them in themselves. Students also learn to listen to others and express themselves understandably and to examine the individual as a group member and the construction of the group's dynamics and sense of community. They also learn to apply their understanding and interactional skills in conflict situations.

Learning and Guidance. The aim of this course is to examine life-long learning and guidance in various situations, and the students consider their own experiences with the help of scientific concepts and to identify central viewpoints in scientific research on the study of learning (different conceptions of learning with their ideas and pedagogical consequences). Students will see the breadth and multiplicity of learning environments and examine learning environments and processes from the learner's viewpoint. The focus is also to analyze the principles and practices of differentiated teaching and the use of different methods to assess learning.

Education, Society, and Change. This course develops students' abilities to recognize how education manifests as a component of social structures and practices and as a force changing these structures and practices. Students examine the cultural, economic, political, and social phenomena associated with the socialization process, as well as educational and training institutions, by using concepts and theories in educational sociology. During the studies, students identify the social phenomena related to childhood, adolescence, adulthood, and the changes that have taken or are taking place in them.

Scientific Thinking and Knowledge. Studies support students to identify the principles underlying everyday thinking and scientific thinking and the differences between them, to examine the basic epistemological and ontological assumptions underlying research, as well as evaluate their own beliefs about knowledge and reality. The aim is also to learn to describe the main methodological lines and various research methods involved in educational research and pick out the central information from research studies and interpret it. Students will construct their own expertise based on research studies and doing research.

Teaching Practice 1. The course focuses on observing and analyzing the activities, learning environment, and pupil diversity within the classroom community from the viewpoint of learning and teaching. Students will reflect on their own experiences and identify their own modes of observation and activity, and recognize the particular pedagogical features involved in learning their own subject. Students plan and implement learning situations and set themselves development goals as educators and teachers. For example besides teaching mechanics students may have special goals for interaction in learning situation. The teaching practice contains, among other things, a minimum of 10 hours of supervised practice and a minimum of five prescheduled periods of a single teacher trainee teaching a full class of pupils. At least one of these five periods has to be a complete lesson, taught independently from beginning to end. Other periods may vary in length and content. For example, three teacher trainees may teach one 45-minute lesson, in which each trainee is responsible for a 15-minute period.

On completion of subject pedagogical studies (35 ECTS), students will have the skills to evaluate their own pedagogical competence and understanding and to set themselves personal learning goals. They will also identify the key characteristics of the learner and learning environment, the risks and available support that affect learning and motivation, as well as the social and structural conditions attached to learning and education. Students will have the skills needed for encountering an individual and a group, for evaluating group processes, and for collegial and multi-professional collaboration. Students will be able to plan, implement, and evaluate learning processes and teaching blocks and to carry out and report a small-scale educational research study.

Subject Pedagogical Studies

Courses in subject pedagogical studies include five modules that are described in more detail in the following section.

Interaction and Cooperation 2. The first goal will deepen students' understanding of themselves as interactors and be able to apply interactional and problem-solving skills related to different encounters. Students will be aware of the relationships of interaction and emotional life to learning and recognize and deal with the ethical phenomena involved in interactional relationships between teacher and pupil and between the group and school community. They will also be able to specify and analyze group processes and an individual's activity as a group member and have an understanding of multiprofessional cooperation in the school community and the skills for supervising and guiding a heterogeneous group.

Education, Society, and Change 2. Since schools are connected to the society, students have to widen their understanding of the trends in social development that affect school, understanding the school community as a culture defining its own activity, and working actively and ethically in diverse and changing communities.

Studies in Educational Administration. This course will give students an understanding of school and education legislation, and students will apply the legislation relevant to the teaching position corresponding to their own level of education.

Learning and Guidance 2. On completion of the course, students will be capable of setting goals for learning their own subject as well as planning, implementing, and evaluating their implementation. Students will be able to focus and individualize learning and educational goals, plan learning units for various lengths of time, and employ appropriate ways of working, teaching methods, and evaluation techniques for different learners.

Constructing Scientific Knowledge: Investigative Teacherhood 2. Since one of the main ideas in Finnish teacher education is to aim for teachers as researchers, this course provides tools for understanding the principles of investigative teacherhood (i.e., teacher as researcher). Students will know the most important qualitative and quantitative approaches to educational and teaching research, become familiar with methods of researching teaching work, learning, and its guidance, and are able to justify the principles and practices underlying their own thinking and actions concerning teaching and education. Students will understand the role of language in constructing meaning and be able to employ pedagogical solutions to support learning, linguistic development, and multilingualism in their own subject.

Teaching Practice

Teaching practice is part of pedagogical studies. The minimum amount of teaching practice is 20 ECTS divided typically into a 2 to 3 year period. At the University of Jyväskylä, the teaching practice begins during the 1st year of studies, but most of it takes places during the 3rd or 4th study year. In the following section, the teaching practice content at the University of Jyväskylä is described.

Supervised Basic Teaching Practice

The course focuses on individualizing the aims and contents of teaching based on the curriculum and nature of the subject. Students will be able to plan and implement teaching units consisting of a few lessons and possess the basic skills for guiding learning and developing their activities based on the assessment and feedback discussions. Students familiarize themselves with the principles of learning assessment and apply this knowledge in their everyday work and are able to make diverse use of various learning environments. During this course, students improve their self-assessment skills and become aware of their responsibilities as educators. The teaching practice also contains 14–16 lessons of individual teaching of a full class and two supervised listening comprehension sessions.

Supervised Advanced Teaching Practice

The course gives tools for planning, implementing, and evaluating learning units that combine pedagogical and theoretical knowledge related to their own subject. Students will be able to guide individual learning, to set goals for their own and their pupils' growth and learning, and to evaluate their own and their pupils' actions in a wide-ranging and versatile way. Students will have an understanding of the diversity of a teacher's work, and they will be capable of developing their pedagogical skills in professional cooperation. Students will recognize and critically evaluate established practices in school culture and apply research knowledge in developing their own teacherhood. Students will be capable of acting as a teacher independently and in cooperation with others. The teaching practice also contains preparing, executing, and reviewing educational lesson plans. A trainee will also have 18–20 lessons of individual teaching of a full class.

Applied Supervised Teaching Practice

The course focuses on the broad perspective and diversity of a teacher's work. Students learn to apply their pedagogical skills in professional cooperation in different work communities and to work in accordance with the professional ethics of a teacher. Students examine their activity as teachers from the viewpoint of the theoretical knowledge affecting education and teaching and understand the importance of lifelong professional growth. They will also be able to examine the teacher's work and school as part of society.

University teacher training schools play an important role in Finnish teacher education (Jyväskylä university training school information: https://www.norssi.jyu.fi/info/university-of-jyvaskyla-teacher-training-school). Training schools are located in all universities that have teacher education programs and are part of the faculty of education. These schools function as typical comprehensive schools, following the national curriculum. Moreover, training schools mentor student teachers during supervised teaching practice. Additional qualifications are expected of teachers, who become experienced supervisors. Many teachers who work at a training school are active in research and development. The teachers working daily at a teacher training school have been educated to act as a tutor for their teacher trainees. Their role as tutors is focused on directing the practical actions and choices that the trainees make at the training school on a given day of teaching their training lessons. In contrast, the didacticians in the teacher training department of the university focus on the development of theoretical viewpoints with the trainees. The training school teachers and the university didacticians collaborate closely on the direction of the overall tutoring and the content of the curriculum of the teacher training program.

Some parts of the practice are performed in schools not connected to the university. These schools are regular schools and represent the everyday practice of schools in general.

TEACHER EDUCATION QUALITY ASSURANCE

Teacher education and teachers in service are not evaluated through a national inspection, by external examiners, or through formal measures. Teacher education quality control is based on department self-evaluation and an international evaluation every 5 years. One particular tool for evaluation is the gathering of feedback from students. It is utilized on a feedback-per-each-course basis after the first university year is over, after a bachelor's degree has been completed, and after a master's degree has been completed. The purpose of the feedback is mainly to develop the content of the courses. The feedback is not included in teacher evaluations. In addition, universities with pedagogical programs have varying internal periods when they collect feedback from alumni members and other connected entities.

There are no national comparisons or evaluations of teacher training. The Finnish Education Evaluation Centre (http://karvi.fi/en/) performs a general university status evaluation every 5 years, but it does not target pedagogical programs in particular. Universities evaluate their educational process in terms of completed degrees, the percentage of advancing students, and the percentage of students who reach 55 ECTS per study year. Because teacher education takes place at scientific universities, the quality of their teaching is also reflected in the research performed by the faculties at said universities. This includes publications that deal with subject-specific pedagogies, which are also a factor in measuring a department's productivity. Today, most faculty members of Finnish education departments have a PhD in either educational sciences or some other field. For example, the lecturers and researchers in the pedagogy of sciences usually have a PhD in natural sciences. In addition, they usually have at least minored in pedagogical studies and completed advanced courses in educational sciences.

FUTURE CHALLENGES IN TEACHER EDUCATION

Reacting to changes in society and schools has generated challenges for teacher education. The ability to use computers and other learning technologies is a key teacher competence. Inclusive education trends also mean that teachers need to have more knowledge and practice in special education. Teachers have traditionally worked alone, but the current changes in working life imply that teachers also have to work in groups, and group work must be provided in teacher education so that future teachers have experience. In our opinion these topics must be included in teacher education, and at the same time, some current topics must be excluded in teacher education programs. Since at the moment in-service education is very minimal in Finland, developing effective in-service politics and implementation is necessary (Evagorou, Dillon, Viiri, & Albe, 2015). In-service education is especially important for teacher students who study pedagogical courses in one year and not as the general trend during the 5 years together with the subject studies.

REFERENCES

Evagorou, M., Dillon, J., Viiri, J., & Albe, V. (2015). Pre-service science teacher preparation in Europe: Comparing pre-service teacher preparation programs in England, France, Finland, and Cyprus. *Journal of Science Teacher Education, 26*(1), 99–115.

Finnish National Board of Education. (2014). Perusopetuksen opetussuunnitelman perusteet_2014 [The core national curriculum 2014]. Retrieved from http://www.oph.fi/download/163777_perusopetuksen_opetussuunnitelman_perusteet_2014.pdf

Basic Education. (n.d.). Retrieved from http://www.oph.fi/english/curricula_and_qualifications/basic_education

Sähköinen säädöskokoelma: 2015. (n.d.). [in Finnish: Finlex Data Bank]. Retrieved from https://www.finlex.fi/fi/laki/kokoelma/2015/

Hakijamäärät 2014–2015 (yhteishaku). (n.d.). Retrieved from https://www.jyu.fi/yliopistopalvelut/tilastot/opiskelijavalinta

Ministry of Education and Culture. (n.d.). *Finnish acts and decrees concerning education.* Retrieved from http://www.minedu.fi/OPM/Koulutus/koulutuspolitiikka/lait_ja_ohjeet/?lang=en

Niemi, H., & Jakku-Sihvonen, R. (2006). Research-based teacher education. In R. Jakku-Sihvonen & H. Niemi (Eds.), *Research-based teacher education in Finland: Reflections by Finnish teacher educators* (pp. 31–50). Turku, Finland: Finnish Educational Research Association.

Niemi, H., Toom, A., & Kallioniemi, A. (Eds.). (2012). *Miracle of education. The principles and practices of teaching and learning in Finnish schools.* Rotterdam, The Netherlands: Sense.

Silander, T., & Välijärvi, J. (2013). The theory and practice of building pedagogical skill in Finnish teacher education. In H. Meyer & A. Benavot (Eds.), *PISA, power, and policy: The emergence of global educational governance* (pp. 77–97). Oxford, England: Symposium Books.

Simola, H. (2005). The Finnish miracle of PISA: Historical and sociological remarks on teaching and teacher education. *Comparative Education, 41*(4), 455–470.

The Bologna Process and the European Higher Education Area. (n.d.). Retrieved from http://ec.europa.eu/education/policy/higher-education/bologna-process_en

CHAPTER 4

EVOLUTION OF THE AIMS OF FRENCH SCIENCE TEACHER TRAINING

Tensions of Implementation

Magali Fuchs-Gallezot
Université Paris-Sud

Maryline Coquidé
Institut Français d'Education

In France, teacher training has been in constant transformation for over 25 years. The latest reform is part of a movement launched in 2012 to rebuild a fair, demanding, and inclusive school (Peillon, 2013). It resulted in the publication of an orientation and programming law for the refoundation of the School of the Republic (Ministère de l'Education Nationale, MEN, 2013a). This law provides a framework that organizes and structures training of all education trades, in particular classroom teaching.

In order to meet the ambition to better train young people, especially the less privileged, to reduce the number of pupils leaving the education

Model Science Teacher Preparation Programs, pages 69–92

system without any qualification, and to better anticipate social, economic, and technological changes, the government affirms how important it is to renew teaching practices and to redefine education trades. In particular, these initiatives aim to recognize broader missions and to give more autonomy to educational teams (Bonneau, Colombani, Forestier, Mons, & Dulot, 2012). Between 2012 and 2018, 40% of teachers currently employed will retire. The training of young teachers is therefore a strong leverage to change the education trades.

This training to the teaching, educational, and training trades is provided by masters named *Métiers de l'Enseignement, de l'Education et et de la Formation* (MEEF: teaching, educational, and training trades) which are strongly framed by the state and which integrate preparation for recruitment competition. These masters are provided by the different *Ecoles Supérieures du Professorat et de l'Education* (ESPE: higher schools of teaching and education). Thirty-two ESPE are spread over the entire French territory (mainland and overseas).

This reform mobilizes three intertwined dynamic processes: a process of universitarisation, a process of professionalization of teachers' training, and a process of unification of the teaching profession (Lebeaume, 2009). The professionalization process goes with the redefinition of educational professions and the recognition of their complexity. It is therefore to form a reflective practitioner (Schön, 1987) able to problematize and critically analyze their practices and those of his colleagues. The process of universitaristion responds for Bourdoncle (1990) to the aim of rationality associated with this professionalism model: This process aims at developing and disseminating intellectual tools to allow distanced and critical analysis of different practices. Finally, the process of unification of teaching and educational professions is visible, for example, in the homogenization of status and remuneration of teachers in primary and secondary level, in the alignment of recruitment levels (master) and in the development for the new reform of a competency framework for education and teaching professions introducing 14 general skills shared by all education and teaching personnel and five general competencies shared by primary and secondary schools' teachers (MEN, 2013b).

The reform carried by the law for the rebuilding of the school of the Republic is based on joint modifications of the competency framework for education and teaching professions (MEN, 2013b), of the recruitment competitions requirements (MEN, 2013c), and of the national framework of the training provided by MEEF Masters (MEN, 2013d) and ESPE projects (MEN, 2013e). These joint modifications are designed to be consistent and convergent (Conférence des Présidents d'Université, 2013).

In this paper, we describe and analyze the impact of this reform on the training of teachers who provide teaching science in secondary schools (Table 4.1).

TABLE 4.1 Structure of Primary and Secondary Education in France

	Type of Establishment	Year or Learning Cycle (Duration)	Pupils' Age	Experimental Sciences' Teaching Is Provided By
Primary Education	Ecole maternelle	Cycle 1 (3 years)	3 to 5	A teacher specialist of primary school
	Elementary school	Cycle 2 (2 years)	6 to 10	
		Cycle 3 (3 years)		
Secondary Education	Collège (lower secondary)	4 years called: 6e, 5e, 4e, and 3e	11 to 14	A professor of SVT (biology and geology) and a professor of PC (physics and chemistry)
	Lycée général (General High School)	3 years called: 2nde, 1ère, and Terminale	15 to 17	

We detail the case of the training provided by the Master MEEF designed to train teachers of *Sciences de la Vie et de la Terre* (SVT: biology and geology) in the Paris-Sud University center, partner of the regional ESPE of Versailles.

After making explicit the national requirements for secondary science teachers' training and their evolution, the national principles of training's organization and the training's contents will be presented and discussed.

NATIONAL REQUIREMENTS FOR SECONDARY SCIENCE TEACHERS' TRAINING

In France, the teaching of experimental sciences in secondary school (lower secondary: *Collège*; upper secondary: *Lycée général*) is supported by two scholar disciplines: Sciences de la Vie et de la Terre or SVT (teaching biology and geology) and *Physique et Chimie* or PC (teaching physics and chemistry). The teaching of each of these school subjects is provided by a specialist teacher, both in college and high school (Table 4.1).

The majority of secondary school teachers are state civil servants, recruited by national and disciplinary competitions. The two French teachers' recruitment competitions are named CAPES (Certificate of Aptitude for Professorship Second degree) and *Agrégation*. They are, then, mutated in schools according to national and regional assignment procedures, in order to meet the schools' needs.

We present an overview of the requirements to be met and of the training paths to be followed to become a secondary science teacher. The discussion is based upon the presentation of historical evolutions of the teachers' training conditions.

Structure of the National Requirements for Secondary Science Teachers' Training

Since September 2013, education staff's training has been provided by the 32 ESPE (higher schools of teaching and education; MEN, 2013a), which are academic components attached to an university. For example, five universities (University of Cergy-Pontoise, University of Paris-Sud, University of Nanterre, University of Versailles Saint-Quentin, and University of Evry) contribute to the ESPE of Versailles. Each ESPE gather all local institutional partners involved in teachers' training: universities, local Education Authority, and primary and secondary schools (MEN, 2013a). The ESPE's teaching staff reflects this diversity of partners. Alongside academics, professionals involved in schools (teaching, inspection, and management staff) and other educational actors (individuals from arts, culture, popular education, and civic education) work in each ESPE.

The *Institut Français de l'Education* (IFE: French Institute of Education), an internal institute of *Ecole Normale Superieure* (ENS) of Lyon, is a national research organization. IFE is working alongside of the 32 regional ESPE, to develop educational research and resources related to teachers' training and their practices' development (Missions, n.d.). It operates through permanent interactions with the educational communities. It aims at developing research on the various forms and practices of education in France and abroad. It also contributes to continuing education for teachers, and spreading scientific resources throughout educational circles. Moreover, IFE has to provide support for the management and assessment of French education policies and on the international scene, to accompany the evolution of educational systems at all levels (primary, secondary, higher education).

Recruitment, Training, and Granting of Tenure

Becoming a school teacher appointed by the state civil service involves satisfying three requirements (see Figure 4.1):

- *A requirement concerning the recruitment:* It is necessary to be a successful candidate in a national disciplinary competitive entrance examination (CAPES or Agrégation) organized by the Ministry of Education which is the future employer.
- *A requirement concerning the training:* It is necessary to validate an internship year (mid-time internship, part-time training provided by the regional ESPE) and to hold a master 2.
- *A requirement concerning the tenure as a civil servant:* It is necessary to get tenure by the Ministry of Education at the end of the internship year.

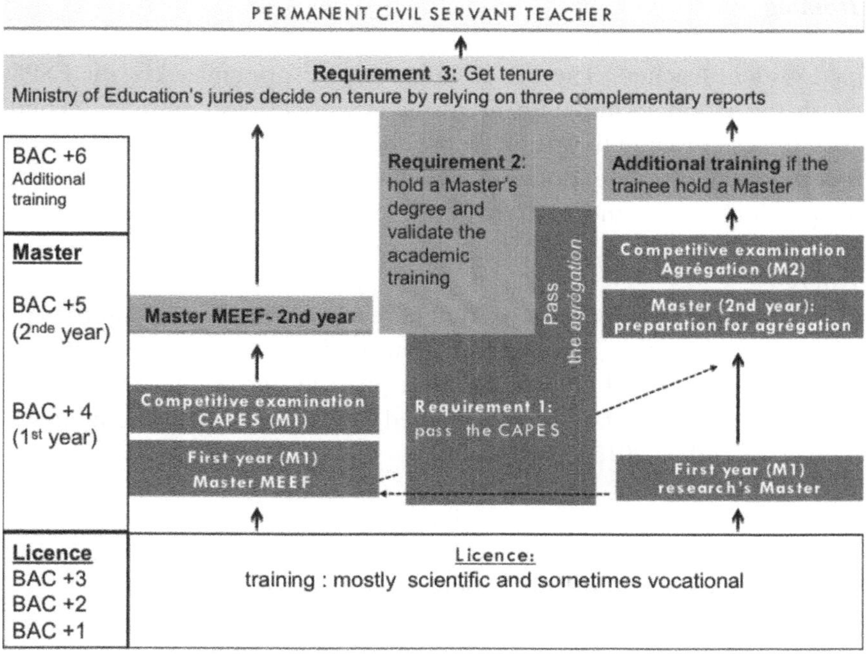

Figure 4.1 General overview of the French teacher training system and of the requirement to become a French permanent civil servant teacher.

Recruitment

Following the completion of a bachelor degree in a scientific field, the students training for careers in teaching mainly follow a master focused on teaching, education, and training careers (called on the entire French territory Master MEEF, see part II). At the end of the first year of the master's degree (M1), they sit the competitive entrance examination, called *Certificat d'Aptitude au Professorat de l'Enseignement du Second degré* or CAPES. To sit the very competitive entrance examination called Agrégation, it is required to follow a second year of a master (M2) dedicated to the specific preparation of this examination. CAPES and Agrégation are both very selective. For example, in 2015, the success rates were respectively 23% for the CAPES for teaching physics and chemistry and 27% for the CAPES for teaching SVT. Only 15% of candidates were successful regarding the agrégation of each of these disciplines (MEN, 2015a).

The candidates who are unsuccessful with CAPES or Agrégation can continue studying in second year of Master MEEF, through an adapted path. After graduation, they apply to posts of temporary or contract staff (untenured teaching posts meeting temporary local needs). Then, an internal competitive entrance examination, taking into account the individual's professional experience, can allow them to obtain tenure as civil servant.

Training

The candidates who are successful with CAPES or Agrégation become paid student teachers. They do an internship year organized by the ESPE, consisting in doing a mid-time internship in school (lower or upper secondary school) and a mid-time training provided by the local ESPE. In the event that the student teacher does not already own a master's degree, the training provided to the student at the ESPE serves as graduation (M2 MEEF).

Tenure

The tenure as civil servant of student teacher is pronounced at the end of the student teacher training year by juries determined by the employer, the French Ministry of Education. These juries decide on tenure by relying on three complementary reports framed by the professional competency framework (MEN, 2015b) (see part II).

- The report given by the head of the school of which the student teacher was in internship which outlines the candidate's successful matriculation at the institutions as well as professional conduct.
- The report of the members of the inspection bodies, which are based, in particular, on the evaluation of the candidate's classroom teaching skills.
- The report of the Director of the ESPE which confirms the trainee's training path at the university. In addition, this recommendation takes into account the obtainment of the master's degree.

History of the National Requirements

Table 4.2 provides an overview of the evolution of different training paths that have been in operation since the establishment of *Instituts Universitaires de Formation des Maîtres*, or IUFM (university institute for teacher training) during Jospin's reforms (1989–1991). Regarding this timeline, we distinguish three periods corresponding to major changes in the organization of teacher training:

- 1989: Creation of University Institutes IUFM which provide teacher training.
- 2010: Development of master including teacher training carried out by the universities,
- 2013: Creation of the 32 ESPE which explicitly take over the training of teachers.

TABLE 4.2 Evolution of the Recruitment and Training of Secondary Experimental Sciences Teachers

		Before 1991	Sessions 1991–2010	Sessions 2011–2013	Since Session 2014
Competitive Entrance Examination	Required level for CAPES	Bachelor degree (equivalent to L3)		Master's degree (M2)	First year of master (M1)
	Required level for Agrégation	Maîtrise (equivalent to M1)		Master's degree (M2)	Master's degree (M2)
	Training for CAPES	Centred on scientific contents (biology, geology for SVT; physics, chemistry for PC)		Scientific (+++) and professional (+) contents	Scientific (++) and Professional (++) contents
	Training for Agrégation			Scientific contents	Scientific contents
	Graduation of the training	The preparation year for these competitive examination doesn't lead to a graduation		Master's degree graduation	CAPES: first year of Master Agrégation: Master's degree
Internship's Year Followed by the Successful Candidates to Entrance Examinations	Type of internship	Full time (1 year)	Mid-time (1 year)	Full time (1 year)	Mid-time (1 year)
	Institution providing professional training	Regional educational centers (Ministry of education)	IUFM (attached to university in 2005)	Regional educational services of the Ministry	ESPE (universities and their partners)
	Support for trainees in the establishment	Educational adviser (expert practitioners)	Educational adviser (expert practitioners)	Educational adviser (expert practitioners)	Establishment tutors (field tutors, expert practitioners)
	Professional training. How many? Who?	Conferences organized by the Ministry	Mid time training provided by IUFM	Short training provided by ministry of education	Mid-time training provided by university (ESPE)
	Graduation of the professional training	No graduation		No graduation	Master's degree graduation
Professional Training	Professional dissertation	No	Yes except for Agrégation	Depending on Masters	Compulsory in second year of Master MEEF
	Position of the professional training in relation to the entrance examination	After the entrance examination		Before	Before and after
		Year in the regional educational centers	Second year of IUFM	During the two years of the Master	During the two years of the Master (including the internship year)

These institutional changes are accompanied by changes concerning the competitive entrance examination, as well as the training and the organization of the internship year leading to get tenure.

We focus the following discussion on the main aspects of these evolutions.

A Rise in the Level of Recruitment and Tenure

During the reform of 1989–1991, the bachelor's degree has become the target level of recruitment for all educational staff. Since 2010, the achievement to a master's degree is required for tenure.

The association of the teacher training at a master's degree graduation has led to recognize the real level of education of teachers, but it has also led (between 2010 and 2013) to a reduction of the number of students sitting the examination of CAPES and Agrégation. Many students were discouraged, especially those from modest backgrounds, by the length of the formation before recruitment and remuneration (Table 4.2), the difficulty of the competitive entrance examination, as well as the low social recognition of education trades (for example a successful laureate to CAPES begins in 2014 with a salary around 1,400 Euros net by month; Auduc, 2012). The repositioning of the competitive entrance examination at the end of the first year of the master helps to restore some attractiveness to educational jobs.

Towards Integration of Professional and Disciplinary Training

The competitive entrance examination and the training form a coherent system. Until the reform of the mastering (2010), the French science teachers had three or four years of scientific education before sitting for CAPES or Agrégation that essentially aimed at assessing scientific content. Professional training took place exclusively during the year of internship, after being successful with one of the competitive examination. At that time, the internship period was frequently the first opportunity for trainee teachers to discover their workplace. Hurried to actually learn their craft, they often requested only advice or "good practices" that could be implemented quickly. In this emergency context, trainee teachers had often granted no legitimacy in the professional contents coming from research on education (Veille Scientifique et Technologique, 2005). Thus the successive training of teachers consisting in scientific training followed by professional training didn't allow an effective professional training.

To promote learning of all contents considered necessary for teachers (educational, pedagogical, didactical, and scientific contents), a simultaneous training for all of these contents has been implemented since 2010. However, between 2010 and 2013, the preservation of a competitive entrance examination assessing mainly scientific contents and the end of the internship year have been detrimental to a strong integration of a professional dimension in the master's teacher training.

Since 2013, the actual reform aims at strengthening the integration of these different teacher training dimensions, scientific and professional. CAPES is now positioned at the end of M1MEEF and designed to assess both scientific and professional dimensions of the training. The year of internship was integrated in the M2 MEEF. Nevertheless, the significant drop in the number of hours of scientific training due to the inclusion of professional training has entertained debates on the decline of the scientific level of candidates to CAPES. Thus, the advocates of a successive formation, scientific then professional (that would begin after success to the competitive entrance examination) and those who think essential to articulate the various dimensions of formation from the beginning of the master's training are opposed to each others.

Affirmation of the Academic Dimension of the Training

Since 2005 and the integration to universities of the IUFM (university institute for teacher training), teacher training is located within the university. This process, known as universitarization, concerns the integration of academic staff and academic knowledge. In particular, the initiation of candidates into education research contributes to build a critical pedagogical stance (Bourdoncle, 1990). One of the challenges of the current reform is the need to set up multi-categorical teams composed of members from the various partners involved in the ESPE: local education authority's services, secondary schools, and universities.

The ESPE pedagogical teams are truly multi-categorical; they are often composed of school teachers, academics in experimental science, and/or in science education, personal management, inspectors, etc. Despite the system's range, an effective and explicit coordination is not always ensured between the various members of the team. The university professors focused on academic areas and on research posture whereas high school staff are more centred on practical and class. The compartmentalized organization of French higher education system restrains collaboration of the various actors and the coordination of the scientific, didactical, pedagogical, and educational content.

Moreover, these professional training modules are new. The academic science education contents are often not mastered by the teaching staff involved as trainers at the university or as tutors in schools. Different ESPE also offer trainers' training masters which aim at training them in science education. A secondary teacher trainers' training program and trainers' qualification has just been announced for 2015–2016. This training is under the responsibility of the Ministry of Education (MEN, 2015b).

NATIONAL PRINCIPLES OF TRAINING ORGANIZATION FOR SECONDARY SCIENCE TEACHER

Based on the recognition that science teaching is a craft that can be learned, the acquisition of professional skills necessary to have the expertise to teach science requires an appropriate professional training which is not limited to mastering science contents. The new teacher training implemented by MEEF Masters was designed in conjunction with the renovation of the requirements for CAPES, and with the professional competencies framework for teaching and educational. Moreover, this framework is used as a reference for training and recruitment.

We present the professional competencies framework, along with the framing of MEEF Masters and of the competitive entrance examination. Then we discuss the impact of this reform on teacher training.

Competency Framework's Features

Teachers' employer, the Ministry of Education, has edited the national standards of teachers' competencies (MEN, 2013b) which is presented Table 4.3. In its preamble, the national standards of teachers' competencies affirms a need to share a common culture between all personnel involved in the public education service, recognizing a specificity of three different professions: primary and secondary teacher, librarians' teacher, and chief supervisor.

Fourteen general teachers and educational staff's competencies neither hierarchical nor sequential, focus on the learning process but also on French cultural tradition of republican and egalitarian values and viewpoints, the relational aspects, and the adaption of its professional practices to changes and the mastery of tools such as ICT or foreign languages.

Five general teachers' competencies are common to primary and secondary teachers independent of their disciplinary specialization. They are oriented on expert practitioners of teaching.

> Within the teaching staff, teachers accompany each student in the construction of his training courses. So that their education promotes and supports the learning process, know-how, and attitudes, they take into account the fundamental concepts related to the development of the child and adolescent, mechanisms of learning and the results of research in these areas. (MEN, 2013b, p. 4)

TABLE 4.3 French National Standards for Teachers

	National Standards	Grouping of Competencies
Common Competencies of Teachers and Educational Staff	1. To share values of the Republic	Concerning regulatory and institutional elements of his professional environment in relation to the responsibilities attached to his function (1;2;6)
	2. To enter action in accordance with the fundamental principles of the education system	
	6. To act as an responsible educator and to follow ethical principles	
	7. To master French language for communication	Relationships, communication and leadership promoting transmission, involvement and cooperation within the educational community and its environment
	10. To cooperate in team	
	11. To contribute to the work of the educational community	
	12. To cooperate with parents	
	13. To cooperate with school partners	
	8. The command of a foreign language	
	9. To master the digital culture elements necessary for the exercise of his profession	Concerning the use and the mastery of IT tools
	14. To engage in individual and collective professional development process	Of analysis and adaptation of professional practice
	3. To know the students and the learning process	Educational and pedagogical competencies necessary for the implementation of learning situations and support diverse students
	4. To take into account students' diversity	
	5. To support students in their training	
Specific Teachers' Competencies	P3. To build, to implement and to facilitate teaching and learning situations, taking into account the diversity of students	
	P4. To organize a mode of promoting group learning and socialization of students	
	P5. To assess progress and achievement of students	
	P1. To master the subject knowledge and its didactic	Related to the mastery of scientific and didactical contents
	P2. To master the French language as part of his teaching	

Competitive Entrance Examination: The New CAPES (Recruitment M1)

To sit CAPES, students must have a first year master level. The competitive entrance examination is now an integrated step in a overall training path. However, it is an act of recruitment and not of certification. It is based on the appreciation of a degree of mastery of scientific and professional skills through the course of acquisition. Thus, for competitive entrance examination for becoming a biology and geology (SVT) teacher, the assessment of dimensions both scientific and professional is covered by the examination. The professional dimension, which was before completely absent of written examination, is now present in one of the written examinations and is the heart of the two oral tests.

Contents' Guidelines

The Master MEEF issued by the ESPE, is defined at the national level by the Ministry of Higher Education and Research (MEN, 2013d). Master MEEF is based, like all French Masters, on two main pillars: specialization in knowledge and skills on the one hand, introduction to research on the other (MEN, 2002). In addition, it includes several types of internship: observation, team-based practice, and responsibility.

The contents of the Master MEEF is built on the national standard of educational and training staff's competencies (MEN, 2013d). Acquiring these skills is ensured through a training that is "progressive and integrated" (MEN, 2013d) which should allow for combining "excellence in one or more academic area(s) and the ability to transmit knowledge" (www.education.gouv.fr, 2013, para. 4). In accordance with the principles of vocational competency construction, a national coordination proposed an elaboration of the master's main structure in module content and assessment by five blocks (Figure 4.2): subject content block (30 European Credit Transfer and accumulation System (ECTS) credits for first year of master (M1), 8 for the second year of the master (M2), didactic block (15 ECTS for M1 and 16 for M2), research block (6 ECTS for M1 and 10 for M2 for the

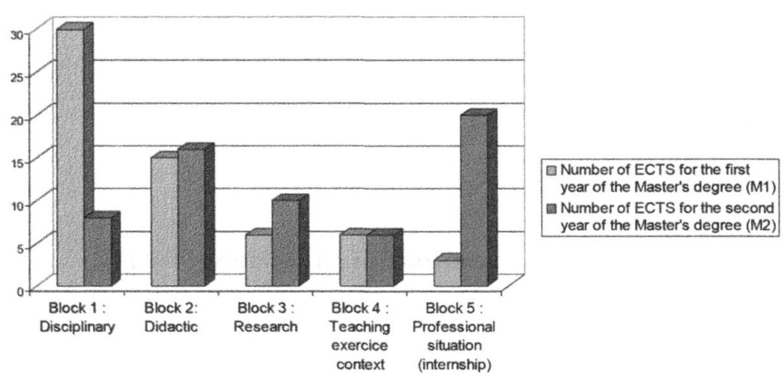

Figure 4.2 National coordination of the Master's degree by blocks.

professional dissertation), teaching exercise context block (6 ECTS for M1 and M2), professional situation block (3 ECTS for M1 and 20 for M2). The M2 is thus centered on professional practice and on the construction of a distanced critical stance while the M1 aims at acquiring the necessary contents for teaching (scientific, didactical, professional knowledge). Content elements in relation to these five blocks are presented in Part III.

Masters MEEF's *Specific Field Experience Requirements*

The decree framing Master MEEF's training claims the articulation of "theoretical and practical courses with one or more internship" (MEN, 2013d, article 1, p. 1). Article 11 of the decree states that "internships contribute to the formation and allow a gradual entry into the profession" (MEN, 2013d). Time for preparation, time for coaching by field and academic tutors, and time for operational and reflective analysis are associated with the various type of internship.

The internship must give at least 20 ECTS during the second year of the MEEF Master.

Various types of internship are recommended (Table 4.4).

- *Observation internship:* Enables the trainee to discover the school and its classes without intervening in the classroom.
- *Practice internship:* Enables the trainee to take over the teachings of one or more levels while being accompanied by a field tutor in the development and implementation of teaching sessions and in the analysis of his practices. The degree of autonomy given to the trainee is left to the discretion of field tutor.
- *Responsibility internship:* The trainee provides teaching sessions standing alone. In these cases, the field tutors' mission is to accompany the trainee.

TABLE 4.4 Internship's Organization

Internship	M1 MEEF		M2 MEEF	
			Laureate	Non laureate
Place in training	S1	S2	Alternation	Alternation
Duration	4 to 6 weeks during the year		Mid-time	8 to 12 weeks
Type of internship	Observation and practice	Team-based practice	Responsability	Team-based practice
Field tutors	The tutor is in the school			
Academic tutors	No visit		2 visits during the year	

Impact of These National Choices on Science Teacher Training

Towards a More Homogeneous Training Throughout the French Territory

In the reform of 2010 (Table 4.3), the absence of a national framework for the masters aiming at training teachers had left ample room for the autonomy of the different universities. Training paths, and particularly the inclusion of professional dimensions, varied greatly from one master to another, depending on universities and disciplines. The variety of these training paths did not give a clear view of which path to follow to become a teacher, which can have contributed in part to the decrease in students' orientation towards teaching. One of the challenges of the actual reform, launched in 2013, is to make the training paths to become a teacher more accessible and readable all across the country. This is one of the reason why this reform is strongly framed by multiple national frameworks.

This strong ministerial framing (Ministry of Education and Ministry of Higher Education) falls in opposition with the autonomy of universities (MESR, 2007) and robs the training teams of a part of the reflection on the development of masters. Indeed, there was little room left by this framing for local elaboration of the masters. However, diversity occurs during the development of masters in the different regional ESPE. This diversity is related to the diversity of partners contributing to each ESPE in the different regions, to local educational history and to the profiles of the local teaching teams' trainers. Hence, all the French Masters MEEF are more or less didactical, scientific, and professional.

The Difficult Implementation of the Common Core Syllabus

One of the challenges of the reform is the development of a common operational and professional culture for all educational staff. The establishment of a common core of training bringing together students training to become primary or secondary teachers regardless of specialty, librarian teacher, or school chief supervisor is seen as a way to develop the common professional competencies made explicit in the national standard competencies framework. However, this conception of the teaching profession combining the teaching-learning assignments and global education (civic, social, or legal) does not go without affecting the current professional identity of French secondary teachers, strongly centered on the teaching of their discipline (Obin, 2013).

For various reasons, these contents have been the subject of a treatment depending mostly on local resources and not on a problematization of the training. The content logic observed was often a matter of juxtaposition, without strong connections to professional skills (IGEN & IGAENR, 2014; Magner, 2014). Institutional partners, like schools, are often cited as needing to be more present in the training offered by the ESPE "so that their innovations and feedback feed the teaching content and methods taught to future teachers" (Magner, 2014, p. 7). Thus, training could be focused from the beginning on the reality of situations coming from schools.

Organizing an Integrative Alternation With Mixed Tutoring

The Masters MEEF display an ambition to set up an integrative alternation in M2: The two training places (school and university) have to co-participate in the acquisition of skills and to share a training project. This implies complementary or joint interventions of the various training actors, shared tools to communicate on the coaching and supporting of the trainees, and the construction of a dynamic relationship based on trust between the trainees, the field tutors, and the academic tutors and trainers. In order to establish a joint tutoring, this double tutoring must overtake the juxtaposition in times and space of the trainee's training at school and at the university and must allow for sharing reflections coming from university training and from school training .

However the deletion of universitary tutors' visits during the internship of the first year of the master do not favor communication between academic trainers and field tutors (IGEN & IGAENR, 2014). The supporting of trainees was not always received as complementary because of the lack of shared tools and because university tutors and field tutors have not enough opportunities to work together. For the year 2015–2016, an intern set of specifications will be tested as a communication tool between trainees, trainers, and tutors. A reflection on the joint assessment arrangements, between university and school, has also begun. Integrative alternation so far remains an ambition. The current reality can be understood as being between juxtaposition of the different trainings and an effort to get them more complementary.

Type and Position of the Professional Dissertation in the Training

The professional dissertation, although presented as an important element of the training, has been completed only by some of the trainees in 2014–2015. For example, in the ESPE of Versailles, all trainees with a master graduation were given the choice between the development, writing and defense of a professional dissertation, or the preparation and drafting of a professional piece of writing. The professional writing, less demanding, shorter and closer to practical analysis was often chosen by the students teachers.

TRAINING CONTENTS FOR SECONDARY SCIENCE TEACHER

As outlined in the previous section, master training content has been defined in a set of constraints: The current reality can be understood as being between juxtaposition of the different training and a effort to get them more complementary.

- Content types and their hourly volume are framed by the expected blocks that must be present in the master.

- The level of expectation of the two years of the master (M1 and M2) are framed by the competitive entrance examination's requirement and by those concerning the professional competencies.
- The nature of the contents (for example, the presence of science education research results) is often dependent on profiles of local actors involved in these training.

The discussion of the contents selected in the master chosen as a case study throughout this chapter is about

- content related to the teaching of biology and geology (SVT) school discipline (educational and scientific content);
- content related to the knowledge of the trade; and
- content related to the acquisition of a reflexive posture.

The theoretical framework of pedagogical content knowledge (PCK; Shulman, 1986, 1987) was not mobilized for the development of training content of this master. It will, however, be used as an post-facto analysis table.

Case of the Training Contents of a Master MEEF for SVT Teachers

The Table 4.5 presents the contents of courses related to biology and geology, the competitive examination, didactics of SVT, educational sciences, practice analysis, an introduction to research, ICT, and a foreign language. These contents refer to the five blocks defined in Part II.

Choice and Impact of Content Related to Teaching a Scientific School Discipline (SVT)

The contents related to teaching the scientific school discipline SVT refer to scientific content, biology and geology, didactics of SVT, and institutional and empirical standards for expected practices in classrooms.

The French Didactics of SVT: A Research Area Taken as a Reference for Thinking SVT Teachers' Training

Martinand characterize the didactic point of view as "the exercise of a recognized and assumed responsibility for educational content" (Martinand, Reuter, & Lebeaume, 2007, p. 107). One key feature of French didactic research focus is to study what has been called the *didactic system* with the aim of a better understanding of how it works. The didactic system is often presented in the

TABLE 4.5 Training Course Content for Teachers of SVT (Case of Master MEEF, ESPE of Versailles, University Paris-Sud)

Contents Related	M1 MEEF (including preparation for the competitive examination) and M2 MEEF designed for the successful candidate to CAPES or Agrégation
Biology and Geology	Block 1 Academic content in biology and geology organized by CAPES' expectations M1: scientific concepts in relation with all subjects of the school subject curriculum M2: enrichment in biology and geology
Competitive Examination	Blocks 2, 3, and 4 Training for written and oral contest (M1)
Didactics of SVT	Blocks 2 and 3 (M1 and M2) *Contents within empirical and institutional registers:* curriculum of SVT; teachings practices (teaching processes, student activities, assessment). *Contents within a more theoretical register:* theoretical research and practical results in connection with class issues (e.g., conceptions of students, science's activities; problematisation, modelisation, nature of science; citizenship education; debating socio-scientific subjects; the use of the history of science)
Professional Practices	Blocks 2, 4 and 5 Assistance in the development and analysis of teaching (M1) Training in practical analysis backed by teaching skills (M2)
Interdisciplinary	Blocks 1, 2, and 4 Knowledge on interdisciplinary, development of an interdisciplinary teaching project (M2)
Introduction to Research	Block 3 Presentation of didactic research results (M1 and M2). Conducting a research in science education oriented towards a professional topic (M2)
Common-Core	Block 4 M1: educational policies, educational sociology, theories of learning. M2: classroom management, teaching posture (voice). Creating a common culture to all education personnel related to learning situations (classroom management, taking into account the diversity of pupils, students with disabilities); to pupils' paths (competency-based approach, specificities of the different levels of education, guidance, learning processes); to the principles and ethics of the educational profession (teaching secularism, gender equality).
ITC	Training in the educational use of digital tools and resources in reference to computer and internet certificate of Higher Education Level 2 for teacher (M1 and M2)
Foreign Language	M2: B2 level of European reference

form of a triangular model presented with the teacher, student, and content as its three corners and the sides as processes such as learning, teaching, and training. Although this model is likely to be limited, it could help to establish more clearly the objectives of the study of didactics (Figure 4.3).

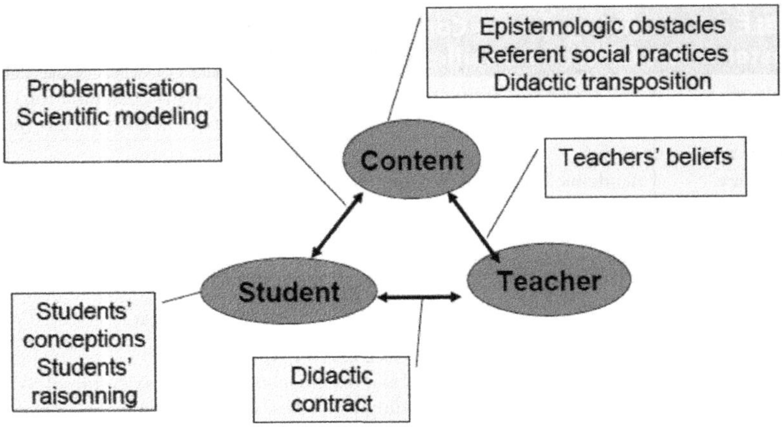

Figure 4.3 The didactic system.

We can also identify some main research tasks types that mobilize researchers in didactic study of disciplines: the interpretation of logical thinking of learners, the analysis of knowledge and taught content, the construction of situations and teaching tools, the contribution of school subjects to transversal education" (Meheut, 2006).

Many of the cited concepts are designed to develop training content. No specific program has so far been set for the CAPES. It is expected that students acquire a critical perspective on the scientific knowledge they will have to teach (MEN, 2013c). In this perspective, content in history of science are, for example, added to biology and geology contents.

From Didactic Research to Didactic Content Taught

Didactic content mobilized for training differs from didactic contents mobilized in research through a change of perspective: The didactic concept abandons the operational function in that it operates within the field of research to be used as a reading grid for class' practices and enrich students' understanding (Astolfi & Develay, 2005). Several limits can be applied to contribution of research. The relationship between theory and practice is not a relation of application but a relation of help to intervention. It should help the future teacher to make more sense of a class activity at the level of the student, of the class, and of a curriculum.

Contents Required for Teaching

By looking at these didactic concepts we can detect the necessity that merely knowing the subject matter is not sufficient at all for teaching the subject. This is the case when considering the PCK perspective (Shulman, 1986, 1987).

PCK integrates different forms of knowledge, beliefs, and values, all of which are essential to the development of professional experience: subject matter knowledge, pedagogical knowledge, curricular knowledge, knowledge of students' understanding (how they understand, their conceptions and usual errors), knowledge of the purposes of teaching, general pedagogical knowledge, and PCK program orientations.

The Relationship Between SMK and PCK

Good subject matter knowledge (SMK) impacts instructional strategies (active/ passive) and give confidence to teachers, but they are not sufficient in and of themselves to have good PCK. It seems that an essential difficulty for teachers' trainees is the complexity of PCK and the resulting difficulty in acquiring it during initial training because of the fact that PCK is encapsulated in the practical actions.

The problem of PCK generalization is also posed. Content knowledge is highly specific to the domain concerned. The question of acquiring a considerable number of PCK remains an obstacle to their integration into teacher education syllabuses. Nevertheless, the question of the explicit link between training in biology and geology and teacher training remains unanswered. Often juxtaposed, these trainings would improve themselves being articulated. For example, a number of interesting links could make this conjunction explicit: bind the study of the biological functions with the study of pupils' conceptions, bind teachers' training to plant photosynthesis with training for experimental practices, and bind geological formation in tectonics with the training in modeling.

Indeed, curricula for science teaching (often focusing on inquiry-based science education) requires reflection on interrelated dimensions associated to science subjects: political and cultural dimensions, scientific dimension, educational dimension, dimension related to curriculum.

Choice and Impact of Content Related to the Knowledge of the Educational Trades

A Design and an Implementation Judged Unconvincing on the Whole Territory

Various institutional reports (IGEN & IGAEN, 2014, 2015; Magner, 2014) point out a fragmented, general training, often limited to a juxtaposition of thematic whose consistency revealed little consideration. These choices appear often arbitrary, sometimes organized primarily according to local human resources. For example, social scientists, inclusive education's specialists are not evenly distributed among all universities. This repartition has impacts on the training which do or do not include these contents.

Nevertheless, the ambition would be for the common-core syllabus to appear as central in the overall training. It calls for a crumbling of borders between the training provided through theoretical courses, internships, and research.

The reaffirmation of the professional dimension of competitive entrance examination, in particular aspects concerning the values of the Republic, is as a tactic to give meaning to this common core during the first year of the master and to contribute to the restoration of values of the Republic in the hearts of the educational staff.

Choice and Impact of Content Related to the Acquisition of a Reflexive Posture

The Analysis of Reflexive Practice

The sandwich education of a trainee teacher combines theoretical and practical training. The elaboration of the teacher training is based on the principles of construction of vocational competence, with three complementary areas (Le Boterf, 2005): (a) *resources axis* based on internal resources such as knowledge, skills, values, identity, or posture and external resources including traditional and digital tools, databases, and actors; b) *activity axis* with different professional situations; and c) *distancing and self-reflection axis* (see Figure 4.4).

Different studies indicate that teachers' good values and belief, good subject matter knowledge, and good abilities to take into account and reflect on feedback influence solid PCK development. Classroom experience seems to be the most influential factor (Kind, 2009), even if it is sometimes not enough. Indeed, PCK development is complex: It relates to trainees abilities to integrate knowledge from a variety of sources. The ability to reflect on feedback seems to contribute to their development. This is what is sought through the implementation of various reflective devices: The reflective practice of analysis and dissertation or professional piece of writing.

The training framework is influenced by Schön's reflexive practitioner (1987). Schön argued that expertise is based, to a large extent, on experienced-based knowledge and on thinking about what is applicable in particular situations. The emerging view of teaching as thought-in-action has repercussions for teacher education in the sense of how to transform knowledge into professional education. Typically, teacher educators have provided student teachers with components of knowledge from a variety of domains of knowledge that can inform teaching. But deciding which knowledge and experience to include raises the question of which concepts of teaching and learning and of the nature of the subject itself underpin the content selection. The prospect of the subject content for the teacher education is not only "to form" but "to form at…something" (such as speaking, considering curriculum, and reaching new professional horizons). It is not just about "knowing what" but also about "knowing how" (e.g., about the subject

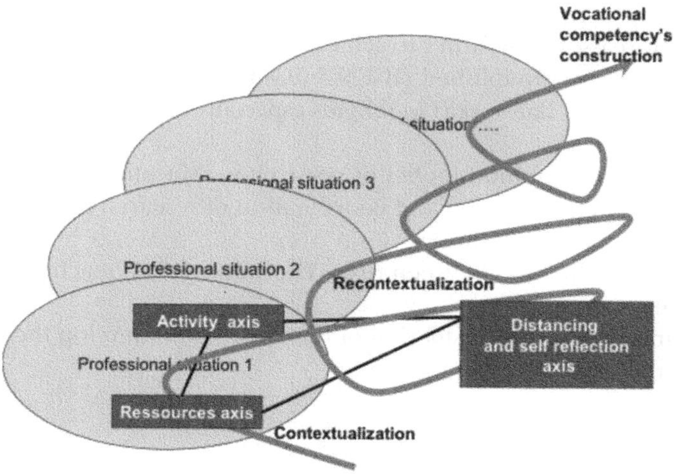

Figure 4.4 Three complementary areas of vocational competencies.

content: to know disciplinary-selected content which could allow teachers to be informed on the identity and historic development of educational issues).

Introduction to Research in Science Education

The aim is for the student to become familiar with various aspects of the scientific research process including research education. Beyond disciplinary content, the research activity should enable the future teacher to develop skills in observation and analysis of his own and his colleagues' professional practices. So, the master contents aim to develop an academic training, with articulation between educational, psychological, epistemological, and curricular records. The master's dissertation is a major element of this introduction to research. We can focus here on the importance of the academic nature of work related to the professional dissertation: requirements of problematization, distancing, discussion, and references.

CONCLUSION

Current teacher training modalities result from recent reforms whose implementation was rapid (e.g., the law on ESPE was enacted in July 2013 and ESPE opened in September 2013). These modalities are the subject of a national framing and a national monitoring by various institutions (e.g., national representation SENATE; Magner, 2014), General inspectorate of education (IGEN & IGAENR, 2014), and by multi categorical monitoring committee. Progressive adjustments are made in each ESPE, according to local implementations choices and their impacts in order to better meet the national specifications. The implementation of this reform is managed by a project management approach: Its impact on the training of teachers will be continually reassessed in order to achieve the goals set by the reform. Thus a report has just been published (IGEN & IGAEN, 2015) on the implementation of the ESPEs. This report highlights especially the necessity to

- initiate an ambitious policy of research in education (recruitment of researchers, valuation and coordination of research in education) and of trainers' training;
- improve the links between research, training, and teaching practices; and
- pursue the national steering of the reform and develop the evaluation tools.

REFERENCES

Astolfi, J-P., & Develay, M. (2005, 6ème ed.). *La didactique des sciences* [Didactics of sciences]. Paris, France: PUF.

Auduc, J-L. (2012). Formation des enseignants: Il faut prendre des mesures tout de suite [Teacher training: Measures must be taken right now]. [site *Le café pédagogique*]. Retrieved from http://www.cafepedagogique.net/lexpresso/Pages/2012/07/16072012Article634780167794001409.aspx

Bonneau, F., Colombani, M.-F., Forestier, C., Mons, N., & Dulot, A. (2012, October 9). *Refondons l'école de la République* [Refound the school of the Republic]. Rapport de la concertation remis au président de la République. Retrieved from http://www.education.gouv.fr/archives/2012/refondonslecole/wp-content/uploads/2012/10/refondons_l_ecole_de_la_republique_rapport_de_la_concertation1.pdf

Bourdoncle, R. (1990). Autour du mot "universitarisation" [Around the word "universitarization"]. *Recherche et Formation, 54,* 135–149.

Conférence des Présidents d'Université, Réseau CEVU, Ministère de l'éducation nationale et Ministère de l'enseignement supérieur et de la recherche. (2013, January 16). *Réforme de la formation des enseignants.* [Compte rendu de réunion CPU-réseau CEVU].

Missions de l'Institut Français d'Education. (n.d.). Retrieved from http://ife.ens-lyon.fr/ife/institut/missions

IGEN, & IGAENR. (2014). *La mise en place des écoles supérieures du professorat et de l'éducation* [The establishment of the higher education and teacher training colleges]. Rapport n°2014–071. Retrieved from http://cache.media.education.gouv.fr/file/2015/45/0/2015-081_Rapport_Final_ESPE_2015_494450.pdf

IGEN, & IGAEN. (2015). *Le suivi de la mise en place des écoles supérieures du professorat et de l'éducation au cours de l'année 2014–2015* [The follow-up of the establishment of the higher education and teaching schools during the year 2014-2015]. Rapport n°2015–081. Retrieved from http://cache.media.education.gouv.fr/file/2015/45/0/2015-081_Rapport_Final_ESPE_2015_494450.pdf

Kind, V. (2009). Pedagogical content knowledge in science education: Perspectives and potential for progress. *Studies in Science Education, 45*(2), 169–204.

Lebeaume, J. (2009). Ce qu'on appelle "formation universitaire." De l'analyse générale au cas de la formation en technologie [What is called "university training". From general case to technology training]. *Recherche et formation, 60,* 39–50.

Le Boterf, G. (2005). *Ingénièrie et évaluation des compétences* [Competency's engineering and assessment]. Paris, France: Editions d'Organisation.

Magner, J-B. (2014). *Rapport Sénat. L'an I des ÉSPÉ: Un chantier structurel* [Senate Report. Year 1 of the SPEs: A Structural project]. Retrieved from http://www.senat.fr/rap/r13-579/r13-5792.html

Martinand, J.-L., Reuter, Y., & Lebeaume, J. (2007). *Contenus, didactiques, disciplines, formation* [Contents, didactics, disciplines, training]. Recherche et formation, n°55. Retrieved from http://ife.ens-lyon.fr/publications/edition-electronique/recherche-et-formation/RR055-08.pdf

Meheut, M. (2006). *L'enseignement des sciences dans les établissements scolaire en Europe. Etat des lieux des politiques et de la recherche* [Science education in schools in Europe: Overview of policies and research]. Eurydice-Commission européenne Retrieved from http://acces.ens-lyon.fr/eedd/climat/recherche/rapport-2006-de-la-commission-europeenne-sur-lenseignement-des-sciences-en-europe/

Ministère de l'Education Nationale (MEN). (2002). *Arrêté du 25 avril 2002 définissant le diplôme national de Master* [Decree of 25 April 2002 defining the National Master Degree]. Retrieved from http://www.legifrance.gouv.fr/affichTexte.do;jsessionid=CA7BCDD998615010AE9A768DBFB31896.tpdjo14v_1?cidTexte=JORFTEXT000000771847&categorieLien=id

Ministère de l'Education Nationale (MEN). (2013a). *Loi n° 2013-595 du 8 juillet 2013 d'orientation et de programmation pour la refondation de l'école de la République* [Law n° 2013-595 of 8 july 2013 for orientation and programming the refoundation of the School of the Republic]. Retrieved from http://www.legifrance.gouv.fr/affichTexte.do;jsessionid=?cidTexte=JORFTEXT000027677984&dateTexte=&oldAction=rechJO&categorieLien=id

Ministère de l'Education Nationale (MEN). (2013b). *Arrêté du 1er juillet 2013 relatif au référentiel des compétences professionnelles des métiers du professorat et de l'éducation* [Decree of 1 July 2013 relating to the competency framework for and education teaching professions]. Retrieved from https://www.legifrance.gouv.fr/affichTexte.do?cidTexte=JORFTEXT000027721614&categorieLien=id

Ministère de l'Education Nationale (MEN). (2013c). *Arrêté du 19 avril 2013 fixant les modalités d'organisation des concours du Certificat d'Aptitude au Professorat du second degré* [Decree of 19 April 2013 laying down the procedures for the organization of the recruitment competitions requirements]. Retrieved from https://www.legifrance.gouv.fr/affichTexte.do?cidTexte=JORFTEXT000027361553&dateTexte=&oldAction=rechJO&categorieLien=id

Ministère de l'Education Nationale (MEN). (2013d). *Arrêté du 27 août 2013 fixant le cadre national des formations dispensées au sein des Masters "Métiers de l'Enseignement, de l'Education et de la Formation"* [Decree of 27 August 2013 setting the national framework for the training provided within the Master's Degrees in Education, Training and Education]. Retrieved from https://www.legifrance.gouv.fr/affichTexte.do?cidTexte=JORFTEXT000027905257

Ministère de l'Education Nationale (MEN). (2013e). *Arrêté du 27 août 2013 fixant les modalités d'accréditation des Ecoles Supérieures du Professorat et de l'Education*

[Decree of 27 August 27 2013 setting the modalities of accreditation of the Higher Schools of Professorship and Education]. Retrieved from https://www.legifrance.gouv.fr/eli/arrete/2013/8/27/ESRS1319423A/jo

Ministère de l'Education Nationale (MEN). (2015a). *Données statistiques des concours du CAPES de la session 2015* [Statistical data of the 2015 CAPES competitions]. Retrieved from http://www.education.gouv.fr/cid83753/concours-du-second-degre-donnees-statistiques-de-la-session-2015.html

Ministère de l'Education Nationale (MEN). (2015b). *Note de service n° 2015-055 du 17–3–2015: Modalités d'évaluation du stage et de titularisation des personnels enseignants et d'éducation de l'enseignement public* [Memorandum No. 2015-055 of 17-3-2015: Methods of evaluation of the internship and tenure of teaching and public education staff]. Retrieved from http://www.education.gouv.fr/pid25535/bulletin_officiel.html?cid_bo=87000

Ministère de l'Education Nationale (MEN). (2015c). *Arrêté du 20–07–2015 relatif à l'organisation du certificat d'aptitude aux fonctions de formateur académique* [Decree of 20-07-2015 relating to the organization of the certificate of aptitude for the functions of academic trainer]. Retrieved from http://www.education.gouv.fr/pid25535/bulletin_officiel.html?cid_bo=91552

MESR (2007, August 10). *Loi n° 2007–1199 du 10 août 2007 relative aux libertés et responsabilités des universités* [Law n° 2007-1199 of 10 August 2007 on the freedoms and responsibilities of universities]. Retrieved from https://www.legifrance.gouv.fr/affichTexte.do?cidTexte=JORFTEXT000000824315

Obin, J-P. (2013). *La condition enseignante dans le second degré et ses récentes évolutions.* [The teaching condition in the second degree and its recent evolutions]. Retrieved from http://www.jpobin.com/pdf6oct2014/2013laconditionenseignante.pdf

Peillon, V. (2013). *Publication au journal officiel de la loi d'orientation et de programmation pour la refondation de l'École de la République. [communiqué de presse]* [Publication in the official gazette of the law of orientation and programming. For the refounding of the School of the Republic. (Press release)]. Retrieved from www.education.gouv.fr/cid72962/publication-au-journal-officiel-de-la-loi-d-orientation-et-de-programmationpour-la-refondation-de-l-ecole-de-la-republique.html

Schön, D. (1987). *Educating the reflective practioner.* San Francisco, CA: Jossey-Bass.

Shulman, L. S. (1986). Those who understand: Knowledge growth in teaching. *Educational Researcher, 15*(2), 4–14.

Shulman, L. S. (1987). Knowledge and teaching. Foundations of the new reform. *Harvard Educational Review, 57*(1), 1–22.

Site education.gouv.fr. (01/07/2013). ESPE : *Une forte dimension professionnelle reconnue par un diplôme de Master* [ESPE: A strong professional dimension recognized by a Master's degree]. Retrieved from http://www.education.gouv.fr/cid72804/espe-une-formation-a-forte-dimension-professionnelle-reconnue-par-un-diplome-de-master.html

VST (Cellule Veille scientifique et technologique). (2005). *La formation des enseignants,* 13. [site IFE] Retrieved from http://ife.ens-lyon.fr/vst/DA-Veille/13-decembre-2005.pdf

CHAPTER 5

SCIENCE TEACHER EDUCATION IN PORTUGAL

From National and Institutional Requirements to the Nature of the Preparation of Teachers

Luís Marques
University of Aveiro

Nilza Costa
University of Aveiro

The views expressed in this chapter, although sustained in the literature, both from research and documental analysis, are the responsibility of the two authors and do not necessarily reflect the perspectives of the authorities of the Portuguese Ministry of Education.

In this chapter, focusing on the Portuguese science teacher education, the authors should highlight the following three points:

Model Science Teacher Preparation Programs, pages 93–127

- The main focus is concerned with science secondary school teachers' preparation, those who teach students between 15 and 18 years of age (grades 10th to 12th).
- The preparation under discussion relates only to biology/geology teachers and physics/chemistry teachers. In the Portuguese curriculum, each one of these pairs of disciplines is taught by the same teacher, except in grade 12 where the disciplines are taught separately (biology, geology, physics, chemistry).
- Our narrative begins in the 70s of the last century, since the 1974 revolution that brought a democratic regime ended with the dictatorship period and therefore has had a significant historical, political social, and educational value.

In a democratic context, education in general, and science education in particular, is so relevant an issue in our society and cannot be approached without a very comprehensive view about the preparation of one of the main actors, the teachers. Following the lead of Portuguese educationalists, we forward the concept of teacher as one who makes learning something to someone (Roldão, 2007), rather than one who transmits knowledge or trains someone to learn skills to someone. Moreover, a teacher is also as a professional who, facing a society that is constantly changing and open to new challenges, needs to be developing continuously (Estrela, 2014).

Although it was desirable to refer to the impact of science teacher preparation nowadays in Portugal, the novelty of this system and the newness of many of its characteristics cause a lack of systematic studies in this area. Therefore, in many situations we will only refer to our experience (as science teacher educators, as a member of the panel of the national agency of accreditation, as a member of the staff teams of programs under accreditation) and advise the need to develop national research to study the impact of the way science teachers are being prepared in our country.

BRIEF HISTORY AND ACTUAL NATIONAL REQUIREMENTS FOR SECONDARY SCIENCE TEACHER PREPARATION IN PORTUGAL

It is mandatory to stress that the revolution of 1974 and the subsequent implementation of democratic social structures have affected most aspects of the Portuguese society. Education in general, was among the main beneficiaries of these changes. Meanwhile, since 1986, Portugal is a European Union (EU) state; therefore, the history of teacher education in this country in recent decades cannot be separated from the Bologna Process (BP), launched in 1999, aiming to create a European higher education area by 2010.

Here, a very short synthesis of this process is presented: It corresponds to a public policy of a meta-state for a university meta-field about a supranational educational policy for all EU membership states aiming a "European higher education space" (Lima, Azevedo, & Catani, 2008). This political decision corresponds to a structure of higher education in three cycles. The first cycle (bachelor), the second cycle (master), and the third cycle (doctorate) lasting three, two, and three years, respectively. The mobility of each student is facilitated through a European Credit Transfer and accumulation System (ECTS). From what is reported previously, the description of teacher education in Portugal is carried out in two phases: before and after the BP.

Before the Bologna Process

Teacher education was provided by higher education institutions (universities and teacher education colleges integrated in polytechnic higher education). The main purpose was to help future teachers develop a reflective attitude towards the improvement of the quality of their practical teaching activities. A set of new public universities was founded by the government in the early 1970s. As far as teacher education was concerned, some of these "new" universities made a different proposal from what was carried out by the "old" ones. Let us see, in short, the outlines of each one of the two proposals.

Within legislation defined by the Ministry of Education (Decree Law nº 443/71), science faculties of the three existing "old" universities (in Oporto, Coimbra, and Lisbon) implemented two new courses lasting 5 years: a degree in physics and chemistry and a degree in biology and geology. Each course was divided into two branches, one called a "scientific" one and another "educational." The curricular disciplines during a period of 3 years were held in common for both branches. At the end of the 3 year period, the students had to make their career choice. Those who wanted to become science teachers chose the educational branch and started then their specific preparation. These courses, the main organization of which is presented in Figure 5.1, emerged as a pioneering and innovative project that triggers a set of changes regarding initial teacher education in Portugal. The project was designed with three different components: the scientific content teaching area, the pedagogical area, and the school practice supervision in a scheme 3 + 1 + 1 (Valente, 2002).

So, this curriculum is organized with the intent to provide future teachers with general and current information related to the main topics they must teach in their subject areas. Meanwhile, the pedagogical knowledge needed in order to carry out well-performed activities both at the school

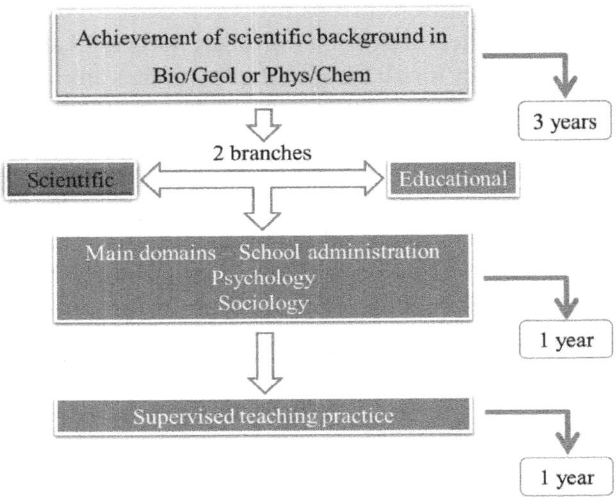

Figure 5.1 Structure of science teachers' education courses at the faculties of science of Portuguese universities before the BP.

in general, and in the classroom in particular, is approached during both semesters of the 4th year.

Finally, during the last year, the students develop their teaching practices in the classroom and participate also in several curricular and non-curricular school activities. The supervision process is carried out by both a science secondary school teacher and a university lecturer, helping in the preparation, implementation, analysis, and discussion of classroom activities and also others developed by the students. At the end of the degree, the students achieve their professional qualification for teaching.

Teacher education was one of the main aims of the "new" Portuguese universities—Algarve, Aveiro, Évora, Minho, and Trás-os-Montes—launched in the late 70s. For achieving this aim, the institutions adopted in their curriculum an integrated model (Campos, 1995), broadly presented in Figure 5.2.

The scientific and educational components of this model are developed, in parallel, along with the curriculum plan, desirably hinged together.

This model was based on the recognition that "teachers' competence is not built by addition but through an integration between academic knowledge, practical knowledge, and transversal knowledge" (Alarcão, Freitas, Ponte, Alarcão, & Tavares, 1997, p. 5), meaning that the preparation of the teachers "must encourage the mobilization and integration of knowledge in context, providing the ability of achieving reality through observation and intervention" (p. 5).

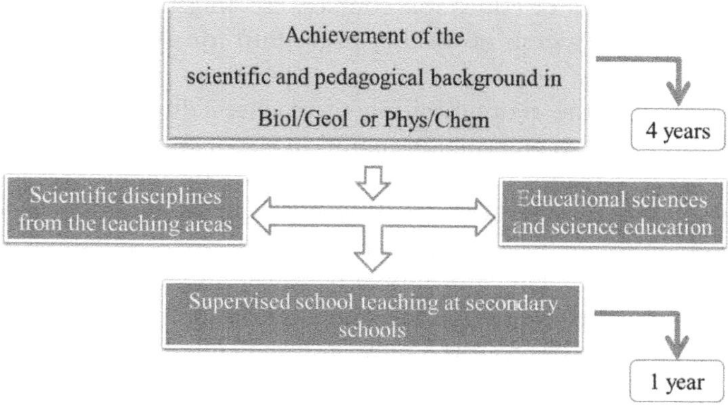

Figure 5.2 Integrated model of the structure of science teachers' education courses.

So, before the BP, initial science teachers education lasted for 5 years and at the end students had a certificate to be a teacher with a degree called *licenciatura.* This could happen in accordance with two different models: a sequential or an integrated model.

The wider process of professional development is at the core of the Portuguese institutions responsible for teachers' preparation, but it is also recognized by in-service teachers. The authors support this view, reporting that since the 1990s several higher education institutions have begun to create academic master's courses and, later on, doctoral courses, which had as target audience teachers in general and science teachers in particular (Cruz, Pombo, & Costa, 2008). Although not specific to science teachers but to Portuguese teachers in general, a national study conducted by the Portuguese Educational Council showed that between the school year 1999/00 and 2009/10, the percentage of secondary school teachers with an academic master and a PhD increased.

After the Bologna Process

In Portugal, legal requirements (Decree-Law 42/2005 and Decree-Law 74/2006, modified by Decree-law 107/2009) are fully filled. The reconfiguration of initial teacher education as part of the reforms implemented by the BP is defined with the publication of Decree-Law 43/2007, which announces important changes for the previous models.

The BP pointed to the structure of higher education courses in teacher education, with a first level of "broadband," focused on the disciplines to be taught, and the second level of expertise in teacher education. Therefore, this process put on the agenda the need to rethink the structure and organization of teacher education courses, creating the opportunity to establish a coherent system of teacher education for all disciplines (Ponte, 2006).

Under the BP, the reorganization of teacher education considers the first cycle of studies as an initial preparation stage, lasting three years, without professional qualification. Initial science teacher education gives now the degree of master. The access to these new courses requires previous academic preparation in the corresponding teaching areas, and this is achieved in 3 years (1st cycle). The educational background takes place mainly during the master course as well as the initial teacher education. This was not the case before the BP—as it was previously explained—which provided an integrated preparation in the teaching area and also in education (Vieira et al., 2013). The legislation (Decree Law 43/2007) highlights that the teaching practice has to be strongly supported by educational research.

"Science teacher education" is related to the preparation and professional development of all those who graduate in two courses: biology and geology teacher education for middle and secondary school and physics and chemistry teacher education for middle and secondary school. Therefore, it is understandable that the structures of the two courses are quite similar. There are a set of common subjects, such as psychology or sociology of education, research methodology, and a group of specific disciplines, for example, didactics or others from the teaching areas (i.e., biology, geology, physics, and chemistry).

Given the massification of secondary education in Portugal after the 1970s, namely, as a consequence of the revolution, the educational system had a strong need of teachers and therefore accepted some without any qualification for teaching (e.g, with an academic background in engineering). In order to solve the lack of teachers' qualification who were already inside the educational system, the Ministry of Education, created in 1988 a law (Decree Law 287/88) which enabled those teachers to get a specialization in teaching (*profissionalização em serviço* [in-service professionalization]). This program lasted for 2 years, with the responsibility belonging to both universities and schools. In the 1st year, the teachers had to do a curricular component at the universities with courses in general education (e.g., psychology and sociology of education), specific didactics, and curricular development. In the second year, the teachers had to develop a pedagogical project in the school under the supervision of two tutors, one from the university and another from the school. This model lasted until now, namely in needed areas, but not in science.

After the BP, the minimum requirement to become a teacher in Portugal is the master. To become a science teacher, every student had to do first a 1st cycle (3 years in a university) with a fixed number of ECTS in the disciplines they will teach in schools. Only then could they enroll in a 2nd cycle (professional master course, 2 years) with a curricular and a teaching practice component. At the same time, with what has been mentioned above, and as a response to the needs to qualified teachers who were already in the educational system, a shorter model, in-service professionalization, existed with 2 years duration.

ACCREDITATION AND NATIONAL REQUIREMENTS FOR SCIENCE TEACHER PREPARATION AND IMPACT IN SCIENCE TEACHER PREPARATION

Before the BP, the licenciatura diploma (5 year course) was required for someone who wants to be a science teacher. This degree, with a grade over 14 out of 20, had also to be achieved by those applicants who wished to undertake the master's and/or doctorate programs, which occurred for a certain percentage of in-service science teachers. For the promotion of quality of teacher education programs, the National Institute for the Accreditation of Teacher Education was created, by Decree-Law in 1998, working as an accreditation body. Legislation more up to date (Decree-Law 22/2014 and Decree-Law 79/2014) has been published, aiming to improve teacher education quality. About the quality assurance the authors note that a professional accreditation of initial education had been developed. This type of accreditation intends to look for whether or not the teacher education program is adequate to the quality demands of professional teaching performance.

After BP, the Agency for Assessment and Accreditation of Higher Education (A3ES) was founded by Decree-Law 369/2007, in line with Law 38/2007 and well-articulated to the European Standards and Guidelines (ESG) endorsed by the Ministers of Education assembled in Bergen in 2005. It is worth stressing that the ESG, endorsed by the Ministers of Education assembled in Bergen in 2005, require that accreditation agencies be independent from governments and institutions (Amaral, Tavares, & Cardoso, 2011).

In practical terms, teacher education institutions design in detail their course proposals and present them to the agency; an international panel of experts analyzes the respective quality and suggest, or not, the approval of the respective accreditation for a period of one, three, or five years. Let's consider, for instance, the case of specific didactics. The study plans of these curricula units have to present the following elements: objectives close to the main science education goals such as enthusing students with a sense of awe and wonder at the natural world; teaching strategies incorporating indications from educational research (i.e., questioning, inquiry bases learning, problem solving); assessment focused on personal development rather than on classification; and updated bibliography both in the teaching and pedagogical areas.

About the "qualification for jobs," it should be noted that quite recently—Decree Law n° 146/2013—a political and very controversial governmental decision introduces a completely new situation related to this issue. Teachers, even after they have completed their preparation in higher education institutions are required to make a written examination, in which they have to reveal their skills for the job performance. We should recognize that it is very doubtful—at least—that this set of pedagogical-didactic issues can be evaluated through a written test!

Given the novelty of this exam, there is not yet studies about its impact; however, it has caused a big controversy among teachers. Also, we do not know research studies focused on the impact of the accretion process in the improvement of the quality of the degrees, and we think this is an area to pursue. Again, this paragraph is an effective summation of the previous subsection and should be tacked on to the preceding paragraph.

STANDARDS REGARDING SCIENCE TEACHER EDUCATION (NATIONAL, STATE, REGIONAL, PREFECTURE, ETC.) AND IMPACT IN SCIENCE TEACHER PREPARATION

In Portugal there are no national or local standards regarding science teacher education, except for a Decree Law/DL (DL n.° 79/2014) which defines the required six scientific areas (general education; specific didactics; cultural, social, and ethical area; educational research methodology; and teaching practice) and the minimum number of ECTS required for each in order to become a science teacher. For instance, it is up to the higher education institutions to define the standards for the preparation of future science teachers. However, given (a) the process of the preparation of secondary school science teachers (a 1st cycle in the scientific disciplines they will be taught and a 2nd cycle in Teaching & Learning in the disciplines to be taught), (b) what is established in the DL n.° 79/2014, and (c) the process of accreditation which is now in development in our country, we believe that there are some common aspects which underscore what a science teacher should be able to know and do at the end of their initial degree. Namely, to know the subjects which they will teach in schools; to know the educational general theories as well as social, psychological, political, and management educational theories and to use this knowledge in different activities (e.g., to prepare pedagogical activities according to pupils' age and psychological development; to interpret educational legislation); to know the didactic principles, methods, and principle areas of research in the specific didactic of the disciplines they will teach; to plan, implement,

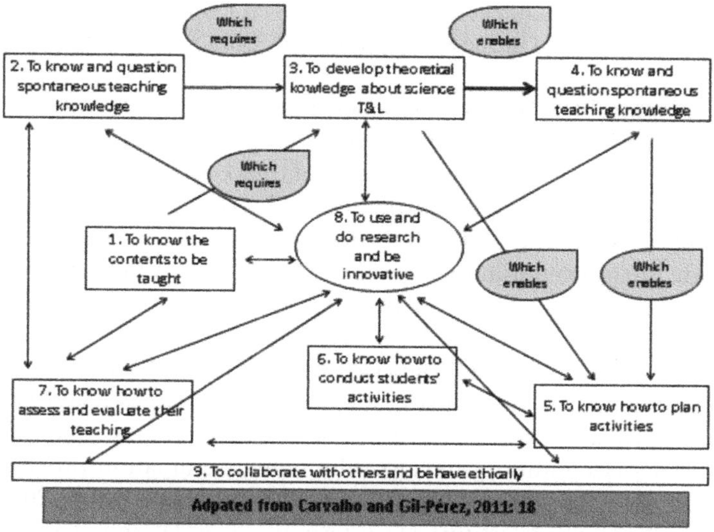

Figure 5.3 Standards underlying the preparation of biology/geology and physics/chemistry teachers at UA in accordance with the experience of the authors of this chapter.

and evaluate teaching practices; to be a reflective student teacher, using research competences; to collaborate with others (with their peers, teachers in schools where they do their teaching practice); and to behave ethically.

This set of aspects is close to the authors' experience at the University of Aveiro, Figure 5.3 illustrates the nine standards which have been underlying their practice.

Also in this topic, we did not have access to any study about the impact concerning the way science teachers are being educated in our country, which is certainly needed.

DIFFERENT LEVELS OF CERTIFICATION AND IMPACT IN SCIENCE TEACHER PREPARATION

Since 2007, the Portuguese government, through its Agency for Assessment and Accreditation of Higher Education (A3ES), which is independent from the Ministry of education, controls all science teacher education courses that are offered by different Portuguese higher education institutions. The main purpose is promoting and ensuring higher education quality.

> The Agency is a private law foundation, established for an indeterminate period of time, with legal status and recognized as being of public utility.

> The Agency is independent in its decisions which must take into account the guidelines prescribed by the State. (About A3ES, n.d.)

A3Es is responsible: to make the assessment of the quality of the higher education institutions courses; to define the accreditation criteria which will translate into qualitative assessments of the courses and the subsequent consequences of the evaluation process; to perform the accreditation of study programs, in order to fulfill the legal conditions for their recognition; to provide the information society regarding the quality of its higher education institutions; and to promote the internationalization of the evaluation process.

The first evaluation of Biology/Geology (B/G) and Physics/Chemistry (F/Q) teacher education courses took place in the school year 2009/2010, aiming to make a screening of that with severe shortcomings and the others with satisfactory conditions to start. The regular evaluation process of the courses under discussion in this chapter was developed in the 2013/2014 school year by panels constituted by international and national experts. The first author participated in some of these evaluation panels. The nature of higher teacher education institutions, profiles of the lecturers involved in the courses, study programs, main scientific areas and ECTS that must be obtained before a degree is awarded, entry requirements, objectives, contents and approached methodologies of the curricular units, characteristics of the initial teaching practice at schools, and profile of school teacher responsible for students' supervision are examples of criteria which have been used. The evaluation results are available at http://www.a3es.pt/pt/acreditacao-e-auditoria/resultados-dos-processos-de-ac

What has just been exposed reveals that a satisfactory quality level of teacher education courses proposed by tertiary institutions is guaranteed by an independent agency.

REQUIREMENTS FOR EACH UNIVERSITY THAT PREPARES SCIENCE TEACHERS AND IMPACT ON NATIONAL REQUIREMENTS, ACCREDITATION, NATIONAL ASSESSMENTS, STANDARDS, AND LEVELS OF CERTIFICATION

The authors have selected three sources that enable the identification of a common denominator related to the set of conditions required by the Portuguese teacher education institutions: teacher education EU guidelines, educational research recommendations, and legal framework.

Four common European principles should be stressed here as far as teacher education is concerned (European Commission, n.d.), based on the assumption that science teachers play a key role in educating learners as EU citizens:

- *A well-qualified profession*: This means that science teacher education is multidisciplinary, the graduation occurs in tertiary institutions.
- *A profession placed within the context of lifelong learning*: Opportunities to go ahead with their studies would be raised, and the "teachers should be encouraged to review evidence of effective practice and engage with current innovation and research in order to keep pace with the evolving knowledge society" (p. 2).
- *A mobile profession*: It should be seen as a possibility of "spend time working or studying in different countries," teaching in different levels of education or, even, in several "professions within the education sector" (p. 3).
- *A profession based on partnerships*: Institutions responsible for teacher education need to ensure that science teaching "benefits from knowledge of current practice" (p. 3) and, therefore, a very well articulation with schools is crucial.

From the research recommendations and taking into account the particular subject of this chapter, the following four items (Osborne & Dilon, 2008) were selected:

- The main science education goal is to help students to understand the material world and also to recognize how science works.
- An innovative attitude related to science curricula is needed. It is important to reformulate objectives, content, teaching and learning strategies, curricular materials, and attitudes in order to strengthen students' motivation for science learning.
- To pay attention to students' needs related to the recognition both of "careers in science" (why working in science is an important activity) or "careers from science (extensive range of potential careers that the study of science affords)" (Osborne & Dilon, 2008, p. 18).
- The relevance of "*extended investigative work and "hands-on" experimentation rather than acquisition of canonical concepts is quite relevant*" (Osborne & Dilon, 2008, p. 19).

Finally, the last source (i.e., the legislation that is produced by the government through the Ministry of Education), Table 5.1 reveals the professional teaching qualifications in the fields of biology/geology or physics/chemistry, achieved through master's degree in the respective specialty (Decree Law 43/2007). The design of these masters courses is an attempt to point out the development of competences through different sorts of activities.

TABLE 5.1 Professional Qualification for Teaching

Teaching Domain	Levels Covered	Master's Specialty	Minimum Credits in the Teaching Area
Biology and Geology	3rd cycle of basic education and secondary education (aged 13–17years old).	Biology and geology teacher education for middle and secondary schools.	120 credits in the sum of the two subject areas none under 50 credits.
Physics and Chemistry	3rd cycle of basic education and secondary education (aged 13–17 years old).	Physics and chemistry teacher education for middle and secondary schools.	120 credits in the sum of the two subject areas; none under 50 credits.

The different competences indicated by the Portuguese educational authorities (DL 241/2001) are organized into three areas (i.e., knowledge, reasoning and communication, and attitudes). A short explanation about each one of these areas follows:

- *Knowledge*: It is related to the achievement, understanding, and use of data, concepts, models, and theories.
- *Reasoning and Communication*: It is concerned with the development of learning and reasoning skills; for instance, researching, analyzing, organizing, critically evaluating, understanding and communicating information, interpreting, critically discussing, judging, deciding, and responsibly acting within the social context.
- *Attitudes*: Here means "the adoption of attitudes and values related to personal and social awareness and to grounded decision-making (concerning problems that involve interactions between science, technology, society, and environment), aimed at an education for citizenship (attitudes and values pertaining to the nature of science and its social implications; rigor, curiosity, humbleness, skepticism, critical analysis reflection, responsibility, cooperation, and solidarity)" (Galvão, Reis, Freire, & Oliveira, 2007, p. 43).

The citizens of our society require well-prepared teachers and, therefore, teacher education institutions—considering the three sources indicated on p. 12—must take into account aspects such as the following:

- Assuming teacher education as a well-qualified profession, to get a master's degree is a *sine qua non* condition to be a science teacher in Portugal, at least since the implementation of the BP.
- In science teacher profession, initial teacher education is not enough and, therefore, all of them are organized to ensure the follow up, for example, at a PhD level.

- As far as mobility is concerned, it has been difficult for our students to attend courses in other countries, even within the EU. It is also unusual to get foreign students enrolled in our teacher education courses. An exception happens with the Portuguese speaking countries.
- Partnerships between higher institutions responsible for teacher education and secondary schools are increasingly important, taking into account the implementation of the teaching practices during the course.
- A student enrolled in the 2nd cycle of the BP or achieving his/her professionalization is required to have previously completed 120 ECTS in the two subject areas—biology and geology, or physics and chemistry—and none of less than 50 (Dec Law nº 43/2007. Curricular units, mainly from a general education area or from the specific didactic area (see Table 5.2), make the mobilization of that scientific knowledge.
- Teacher education institutions should recognize that it is desirable, for the purposes and quality of the courses, to maintain a good articulation with the employer's requirements (mainly secondary schools). Teaching strategies have to contribute to the development of demanding professionals engaged, for example, in lifelong learning. Therefore, the institutions need to monitor the training processes through the regular administration of written questionnaires to the students, usually at the end of each semester. The results obtained are under analysis, usually by the board of the course, and they are even discussed with teachers and students to support any adjustments aiming towards the maintenance of students' motivation for the teaching profession.
- Institutions should assume the importance of the performance of science teaching education courses, and so, the relevance of including particular contents in the study plans that help the learners to achieve an investigative attitude in the context of their professional life.

The analyses of those plans in the seven institutions reveal that in all of them, one can find that, more or less explicitly, subject matters such as

TABLE 5.2 Compulsory Distribution of ECTS for the Different Scientific Area

Scientific Areas	Number of ECTS
General Education Area	minimum 18
Specific Didactics	minimum 30
Teaching Area	minimum 18
Introduction to Professional Practice	minimum 42

science teaching and its dimensions—epistemological, psychological, and sociological—practical work, sustainability and science education, problem solving, education in a science–technology–society context, assessment, inquiry-base learning, youth psychological development, and research methodology, are under discussion. A cross and articulated approach of this set of issues during initial teacher education (constantly in the spotlight regarding its quality and effectiveness) is a way to fit with international recommendations (Rocard, 2007, p. 17). Such recommendations require accurate curricular approaches fitting with the reflexive teacher's continued development (Schon, 1983).

CONTENT REQUIREMENTS (NUMBER OF HOURS, NUMBER OF COURSES, NUMBER OF UNITS, ETC.) FOR DISCIPLINE OR CONTENT AREA LICENSE (e.g., CERTIFICATION OR LICENSURE FOR A PHYSICS TEACHER, BIOLOGY TEACHER, ETC.)

As previously referred to, the 1st cycle ensures in-depth preparation in the teaching areas. All the curricular units address scientific issues; in the case of teacher of Phy/Chem: mainly physics and chemistry and also, maths. Biol/Geol, mainly biology and geology and also basic maths, physics, and chemistry.

The 2nd cycle complements the previous preparation and is mainly concerned with the knowledge acquisition needed for the teaching areas: specific didactics, general education area, teaching areas, and also introduction to professional practices with supervision at schools. The number of ECTS needed to get the master's degree is 120, distributed as described in Table 5.2, for the different areas.

As far as the organization of the two master's course is concerned—biology and geology teacher education for middle and secondary schools, and physics and chemistry teacher education for middle and secondary schools—there are similarities already explained. Therefore, the authors have decided to present, in Table 5.3 and Table 5.4, an overview of the study plans correspondent to the same course: biology/geology teacher education for middle and secondary schools carried out in two higher institutions.

The selected two institutions fulfill a set of characteristics placing each one in opposite positions within a range of several possibilities.

The following criteria have been highlighted in order to reach this distribution: (a) importance given by a particular institution to each one of the scientific areas (in fact, the same area can, in each institution, be assigned by a different number of ECTS), in accordance with the current legislation;

TABLE 5.3 Overview of the Study Plan Highlighting the Importance of the Options

Scientific Areas	Curricular Unit	Duration	Working Hours	ECTS
General Education component (18 ECTS)	Curriculum & Assessment School & Society Educational Process Development & Learning *General Education Component *Education & Media *Citizenship Education *Health Education *Adult Education & Training *History of Education in Portugal *Indiscipline & Violence at School *Special Educational Needs	1 S	168	6
Teaching component B/G (18 ECTS)	*Subject Teaching Component *Geological Risks *Hydrogeology *Other options chosen by the student depending on the individual needs	1 S	168	6
Specific didactics component B/G (30 ECTS)	Didactics of B & G I Didactics of B & G II Methodology of B & G I Methodology of B & G II Seminar of Didactics of B & G	1 S	168	6
Introduction to Professional Practice B/G (48 ECTS)	Introduction to Professional Practice I	1 S	168	6
	Introduction to Professional Practice II		168	6
	Introduction to Professional Practice III		336	12
	Introduction to Professional Practice IV		672	24

* Optional subjects

(b) the weight of the teaching area, reflecting the culture and the nature of the institution itself, is clearly stronger than in the other one; (c) the recognition of the educational relevance of the optional subjects in the study plan is clearly assumed in one of the institutions, while in the other they are completely excluded; and (d) the role played by science educational research staff is highlighted in the institution with a wide range of options.

General education area: the set of disciplines integrating this area vary from one institution to another, whilst trying to achieve the common purpose, in the context of each study plan. The list of options indicated in Table 5.3 includes most parts of them. Meanwhile, in some institutions, one

TABLE 5.4 Overview of the Study Plan That Very Much Emphasizes the Scientific Teaching Area

ScientificAreas	Curricular Unit	Duration	Working Hours	ECTS
General Education (18 ECTS)	School Organization and Classroom Management	1S	108	4
	Psychology of Development & Learning		135	5
	Special Education Needs		108	4
	Curriculum Development & Evaluation		135	5
Teaching Area B/G (24 ECTS)	Observation & Experimentation in B	1S	162	6
	Observation & Experimentation in G			
	Educational Research Project in B & G I			
	Educational Research Project in B & G II			
Specific Didactics B/G (30 ECTS)	Didactic of Biology I	1S	202.5	7.5
	Didactic of Biology II			
	Didactic of Geology I			
	Didactic of Geology II			
Introduction to Professional Practice B/G (48 ECTS)	Pedagogical Training & Report	A	648	48

can identify some more, such as: educational technology, leadership and school management, and ethics in education. The number of ECTS related to each of them, ranges between 3 and 6.

- *Teaching area*: The number of ECTS that in each institution is assigned to this scientific career is an indicator of the science teacher education culture of the respective institution. The study plans for the most part of these institutions (four out of seven) give to this area a minimum number ECTS, (i.e., 18). The authors mention that during the 1st cycle, the corresponding 120 ECTS are fully achieved within the teaching areas. Therefore, it is understandable that the main aim of the 2nd cycle is to complement the preparation in those areas, addressing issues such as:
  - In biology/geology: geological resources, geological hazards, ecological footprint and use of raw material, evolution, importance and threats to biodiversity, direct and indirect factors in the destruction of habitats and species, sustainable development,...
  - In physics and chemistry: modern and quantum physics, astronomy, nanotechnology; polymers, food chemistry and the environment, chemistry and sustainability...

- *Specific didactics*: the nature and the design of the disciplines that incorporate this area is quite similar (name, duration, main aims, working hours, ...) in all institutions, but one. Table 5.3 reveals that the scientific area of one of the institutions consists of two types of subjects: didactic and methodology. The first one is concerned with a set of theoretical issues such as the subject matter under analysis, main aims, methodologies, strategies approaches diversity, and assessment, while the latter is mainly related to the planning and implementation of curricular contents taught in the classroom.

  One of the main concerns in all the institutions is that the designing of this set of curricular units (i.e., specific didactics), is close to the concerns and recommendations of several international reports, for example (Rocard, 2007), "because Europe's future is at stake decision-makers must demand action on improving science education from the bodies responsible for implementing change at local, regional, national and European level" (p. 17).
- *Introduction to professional practice*: No doubt about the relevance of this area when a considerable percentage of credits—40%—is assigned, by law, as it is presented in Table 5.2. Recognizing that "telling is not teaching and listening is not learning" or using a particular set of activities in the classroom, and thinking that "the ideal teacher understands how students learn and recognize a number of factors that impinge on the quality of students' learning; and, on the basis of that understanding, chooses and employs teaching procedures and approaches to promote learning" (Loughran et al., 2012, p. 6). One can realize that teaching is really problematic and very much needs an accurate supervision from the beginning. A set of procedures for the initial teaching period are planned at the institutional level. From the authors' view, the following should be stressed: design of educational class plans; teacher training in specific disciplines—according to each of the two masters course—nature science, biology and geology, or physics and chemistry; intervention in school environment, looking for a progressive and sustained improvement of the learning professional knowledge; and design and writing of a teacher training report in parallel with the development of a portfolio: an appropriate tool for evaluation of the teacher professional development (Sá-Chaves, 2005).

A few words about a very relevant and compulsory activity (i.e., the design, implementation, and evaluation of a small educational research project). The learning with which the students are more familiar is that which is coming "from words," rather than that "from experience" (Russell, 1993) and therefore, it would be beneficial to help the future science teachers to feed his/her reflection, the achievement of an important aim: to develop a research attitude. The subject matter of the investigation is usually centered on issues relating to the classroom as an attempt to diminish the gap between educational research and teaching practice.

GENERAL PEDAGOGICAL COURSES REQUIRED FOR SCIENCE TEACHER PREPARATION

The presentation of the pedagogical courses provided by each one of the seven Portuguese higher institutions responsible for science teacher education are presented in Table 5.5. The several courses presented are affected by each one of the scientific careers already mentioned, although there are some which are almost common to all, like the core disciplines of educational sciences (e.g., sociology of education or psychology of education).

The information here provided is complemented by additional data to be presented in Table 5.6.

CONTENT PEDAGOGICAL COURSES REQUIRED FOR SCIENCE TEACHER PREPARATION

It has already been explained that the students are faced with pedagogical courses (i.e., curricular units, terminology of the BP), only at the 2nd cycle (see Table 5.5). There are in the various institutions curricular units labeled with the same name, or quite similar. For instance, sociology of education, psychology of education, and research methodology. Table 5.6 presents curricular domains contributing to the integral development of students and addressed through school age. Each one includes curricular units, interconnected and taught within the various institutions. Finally, a short synthesis of the correspondent contents is put forward. As can be seen, the content pedagogical courses required are from the educational sciences (sociology, psychology, history and philosophy of education, politics and management in education), but also from didactics and ICT, curriculum development, research methodology, teaching practice, and supervision. The future teachers are faced with general content knowledge about the fundaments of education, the educational system organization, the curriculum in schools, classroom management, didactical principles, and methodologies (namely, sciences and educational research methodology).

A note for emphasis: Some important issues such as special educational needs are explicitly stated in a quite small number of study plans. By personal communication and also from their own experience, the authors have obtained information indicating that this content is often discussed in

TABLE 5.5 Pedagogical Subjects Taught in Each One of the Teacher Education Institutions

Institution	General Education Area	Teach.Compon	Sp. Did. Comp.	Int. Prof. Practi
Institution 1	Sociology of Education and Teaching Profession Guidelines of Pedagogy Psychology of Learning & Motivation Educational Technology *School Organization *Psychology of Adolescence	Physics and Chem of the Universe Physics & Chem in Action: From the Real to the Virtual Seminar in Biology Seminar in Geology	Methodology of . . .** Design and Evaluation of Curricular Materials in . . .**	Educational Coordination and Class Direction Professional Internship
Institution 2	Psychology of Development Multimedia and Scientific Education for Sustainability Education and School Organization Option*	Complementary of . . .** Option***	Curricular Development and assessment Didactic of . . . I & II**	Teaching Practice including Supervised Teaching Project
Institution 3	History and Theory of Education Psychology of Education Education Policies and Educational Organization Educational Research Educational Themes and Problems	ICT applied to Physics ICT applied to Chemistry Research Seminar in Physics Research Seminar in Chemistry	Didactic of Science Didactic of Physics Didactic of Chemistry	Professional Internship in Physics and Chemistry
Institution 4	Psychology of Learning and Development Sociology of Education ICT in Science Organization and School Management Observation and Assessment of Teaching Practices	Geology, Society and Environment Biology, Environment and Society	Curricular Development and Didactic of B and G I and II Didactic Seminar of B/G	Supervised Teaching Practice
Institution 5	Sociology of Education Organization and School Management ICT and Education Special Education Educational Research	Sedimentology** Environmental Geochemistry** Geomorphology** Environmental Management** Biology Applied to Teaching**	Didactic of Geology Didactic of Biology Curricular Development	Interdisciplinary Seminar I and II Professional Internship I and II
Institution 6	Information in Table 3			
Institution 7	Information in Table 4			

* Optional curricular unit; ** Physics and Chemistry/Biology and Geology; *** option to offset shortfalls of ECTS in the1st Cycle

TABLE 5.6 Main Contents of Pedagogical Curricular Units Taught in Science Teacher Education Institutions

Curricular Domain	Curricular Units	Synthesis of Contents
Sociology of education	Sociology of Education Sociology of Education and Professional Teaching Education and Society	Subject matter of sociology of education and contribution for professional development, Globalization, citizenship, inclusion and exclusion social school dimensions; Curriculum social perspective; Nowadays school challenges; Problems of indiscipline and violence
Psychology of education	Psychology of Development and Learning Psychology of Education Educational Process Development and Learning Psychology of Motivation and Learning Psychology of Learning and Educational Value	Human development and theories of human development; Psychological perspectives about the development throughout life Learning and learning theories; Pedagogical relationship as a singular and diversified relationship: factors and dimensions; Educational relationship and pedagogical relationship. Development and learning in school context: interactions and management in the classroom, indiscipline; Educational needs and individual differences; Learning in the future in an ICT context
ICT and education	ICT and Science Education ICT applied to Physics and Chemistry Educational Technology Communication and Education Technologies Multimedia and Science Education* Technology and Innovation in Education Education and Media* New Technologies in Bio/Geol Teaching*	Potential of ICT in Science Education and in promotion of scientific literacy Concepts of multimedia, hypermedia Evolutionary transition of Web1.0 to Web 2.0 Collaborative construction of documents: Google Docs; Wikies

(continued)

TABLE 5.6 Main Contents of Pedagogical Curricular Units Taught in Science Teacher Education Institutions (continued)

Curricular Domain	Curricular Units	Synthesis of Contents
Didactic, Curriculum Development and Special Education Needs	Didactics and Curricular Development in Bio/Geol or Phys/Chem Didactic of Biol or Geol or Phy or Chem Methodology of Biol/Geol or Phys/Che Design and Assessment of Curriculum Materials Curricular Development Special Educational Needs	Concepts of Didactic and Curriculum: epistemological and psychological foundations. Science teaching methodologies: lecturing, scientific inquiry, conceptual change, inquiry-based teaching Scien-Techn-Soc-Envi (STSE) approach Planning: references to specific teaching strategies in Bio, Geol, Phys or Chem. Problem based learning; Questioning. Lab and field work activities Students' learning assessment
Research Methodology	Educational Research Methodology Educational Research Research Seminar	Purpose of the educational methodology Qualitative, quantitative, and mix methodologies: research tools. Analysis of case-study, action-research,. Data analyses –content analysis, SPSS
Educational Policies, Organization and School. Administration	Educational Policies & School Organization School Organization Leadership and School Management Organization School Management	Analysis of the school organizatio Image of the school organization School organizational analysis Organizational image of the school Differently pes of school planning School and educational administration organization. Management of education systems
History, Philosophy of Science, Ethics	History and Theory of Education History of Education in Portugal History and Philosophy of Education Ethics and Deontology in Educational Practice*	History of Education: roots of pedagogical thinking and relevance for science teachers; From Faure and Delors' Reports to a sustainable development Ethicsandprofessionalidentity
Introduction to Professional Practice	Class Management & Educational Supervision Teaching practices Research report	As defined by law the professional traineeship corresponds to; classroom and school activities, under the supervision of cooperative teachers and academics; report of an individual research project subject to public discussion by a jury

* Optional curricular unit; ** Physics and Chemistry/Biology and Geology; *** option to offset shortfalls of ECTS in the1st Cycle

seminars to help students' future teachers overcome some of the difficulties they eventually encounter in school teaching.

SPECIFIC FIELD OF REQUIREMENTS (HOURS IN THE CLASSROOM, ROLES PLAYED BY STUDENTS, ...) BOTH OFFICIAL AND CUSTOMARY

Despite "using evidence about learning, teaching and schooling is not a straightforward matter of applying research findings to teaching" (Florian and Pantic, 2013, p. 6). There is no doubt about the research relevance related to improving the quality of teachers (pre- and in-service) and, therefore—considering the opinion of several authors, for instances Costa, Graça, & Marques (2003)—to provide high-achieving systems of schooling. Several studies (e.g., Furlong, 2013) suggest that improving the quality of teacher education is particularly relevant to ameliorate the teachers' competences. In the context of this section, it is understandable to focus on access and engagement with scholarly and disciplinary literature, as well as knowledge and participation about scholarly communities (Marques, Loureiro, & Marques, 2015). Finally, the BP established a new perspective related to the initial teaching practice. This legislation strengthens the relationship between teaching and research and also implies the obligation of producing a report to be discussed in public examinations (Vieira et al., 2013).

Still, in accordance with the Portuguese law, professional qualification has to be acquired in the context of formal partnership established between the tertiary institution responsible for teacher education and the secondary school. The framework of the supervised students' teaching practice is defined by the following four rules: (a) observation and cooperation in situations of teaching, in particular, and education in general. Involvement in supervised teaching activities in the classroom, within the school context; (b) experience of planning, teaching, and assessment in accordance with the teachers' assignments and competences inside and outside the classroom; (c) teaching activities taking place at different levels for those the respective science teaching course prepares; and (d) all activities are designed in a professional development perspective, targeting a critical and reflexive view about the ethical and civic dimensions of the teaching activity.

A school community is a main concern for both educators and legislators for enabling the students/future-teachers to achieve a satisfactory educational experience. Therefore, institutions responsible for science teacher education are required to make cooperative protocols with secondary schools that are available to work with trainees' groups. The selection of the cooperative teachers (i.e., the teachers who are in charge of the follow-up of the students) in all the activities integrating initial teaching practice, is a very important issue. Science teacher education institutions are responsible for that selection. Nevertheless, the school board should express their agreement as well as the teacher him/herself. The cooperative teachers' profile requires the following: scientific, pedagogical, and supervision competences; teaching experience over five years; and post-graduation in supervision. The last is not mandatory, however, it is very desirable.

A final comment about the way the assessment of this important scientific area (see Table 5.2) is approached. Students' assessment is focused mainly on aspects of the classroom and the school, in a personal and reflexive way (through a portfolio, for example), and in a research dimension (research report). Although the students have a mark at the end (in a scale 0–20), the assessment has a continuous dimension in the sense that the tutors are expected to give to students feedback about the way they are doing and performing in their teaching practice (for example in weekly meetings) or in the development of their research report (with the supervisor). It is expected that the assessment process is negotiated by the actors involved internally (tutors and students), and that criteria and instruments (for instance, assessment grids) are developed together. On the bases of such criteria and instruments, students usually do their self and peer assessment. The final assessment of the research report is completed by an external evaluator on the basis of a grid defined by the higher education institution.

In synthesis, Table 5.7 presents what can be labeled as the identity card of the internship carried out at school. A few comments about the implementation of the initial teaching practice. The amount of time spent at school by the students changes from one teacher education institution to another. For instance, at institution 6 (see Table 5.3 and Table 5.5) the students/future-teachers have activities to fulfill at school from the beginning (i.e., the 1st semester of the course). As one can expect, the relevance of the assigned activities progressively increases, as it is revealed by the number of ECTS awarded in each of the four semesters 3, 6, 9, and 30, respectively. The main goals of each semester are as follows: to know the school context through observation and interviews with several school elements (semester 1); to assume limited responsibilities related to teaching situations (e.g., planning, implementation, evaluation, and class direction; semester 2); to perform supervised teaching practices, within the supervisor classes, for 4/6 weeks, setting a problem concerning to the teaching area, that will be under research through the next semester (semester 3); and to develop the research project under the guidance of the university supervisor and prepare the respective report to be defended in public examination (semester 4).

Institution 4 (see Table 5.5), among others, has a different procedure. The students don´t visit schools during the 1st semester, and if they go there

TABLE 5.7 Main Characteristics and Details Related to the Initial Teaching Practice

Items	Description
Position in the course study plan	Curricular unit of the 2nd year.
Duration	Carried out during 1 year.
Local	Secondary school.
Matter	All the activities performed by the student in the school context, under the supervisors' responsibility.
Purpose	To develop knowledge, skills and attitudes, relevant for teaching, in a school context. To in depth the articulation between teaching and research. To look for a progressive and supervised students' integration in the school community.
Students' tasks	Participation in the planning of teaching activities that will be implement in the classroom. Assistance to the cooperative teacher's classes. Supervised teaching practice 4/6 weeks, in classes assigned to the cooperative teacher. Participation in school and university meetings, related to: the design and evaluation of teaching activities, the scientific and pedagogical issues and the foster the school dynamic. To design, implement and writing a short research project, that will be presented to a jury. Participation on their own assessment.
Cooperative teachers' tasks	Scheduling the teaching practice period for each student in his/her own classes. Ensuring weekly pedagogical supervision of the trainees' teaching practice, mainly the planning preparation, class attendance and subsequent critical analysis. Participation in meetings, at school and university, in discussions with university tutors and foster the integration of students in the school community. Assessment of the trainees, together with the university tutors. Teaching.
University tutors' tasks	To ensure the supervision, particularly about the scientific and educational issues,, in articulation with the work plan. Running sessions aiming to overcome difficulties find out in the classroom. Supervision of the research project. Assessment of the trainees in articulation with the cooperative teacher.
Outputs	Final report/portfolio related to development of teaching practice. Research report to be presented and discussed before a jury.

in the 2nd semester—in some institutions—it is to develop a set of tasks, such as class observation, as curricular activities of specific curricular units are integrated in the course study plan.

In the second year, students/future-teachers develop, at schools, the majority of their curricular activities (usually each school receives 3–4 students). Each student teaches in the supervisor classes for 5 to 6 weeks, usually over two periods of time. As far as supervision is concerned, one notes that the student follow-up by the cooperative teacher is permanent, but the university supervisor's contribution also happens. The role of the latter is especially important in the design and development of the research project, which must be linked to problems emerging from issues that occur in the classroom (see outputs in Table 5.7).

The school plan activities of each school, often linked to the local community, have a huge influence on the tasks performed by the student outside of the classroom. Very often, the dynamics of this plan are carried out by students/future-teachers, for example sessions about local environment issues to the community. Here is one of the reasons why schools reveal willingness to accept the internship groups.

AN ANALYSIS OF THE STRENGTHS AND/OR AREAS FOR IMPROVEMENTS OF SCIENCE TEACHER PREPARATION

Content Knowledge (CK)

Several authors, (e.g., Hannah and Lisa, 2008), refer to the importance of CK for good teaching. The requisite of a significant number of ECTS in the disciplines students will be taught, as seen in Tables 5.2 and 5.3, in initial teacher education in Portugal, is an existing strength of our system. From the Decree Law 79/2014, the proper academic preparation for the teaching demands of the respective scientific area is reinforced, in comparison with the guidelines provided by the Dec. Law 43/2007. This strengthening of the scientific component happens even during the 2nd cycle of the BP. However, given the nature of scientific knowledge (a human kind enterprise, dynamic, holistic, . . .), the fact that it isn't restrained to a body of knowledge but also relates to the methods of its development, the fragmentary curricular approach, a focus on pure knowledge and a passive teaching and learning method, is opposite to an active one. Yet, it is still much in use in Portuguese higher education institutions despite challenges (Costa, 2015). Based on the literature, for instance, Pedrinaci et al. (2013), King (2012), Bozelli and Nardi (2012), Oliveira and Oliveira (2013), and in the authors' experience, we illustrate two examples, one for a course in geology and another one for physics with different focuses: the 1st more focused in the core of the scientific knowledge and its evolution and the 2nd on the teaching and learning methodology.

Geology

The understanding of how the Planet works should be, in the authors' opinion, one of the main goals of compulsory education. Assuming that the contents are tools for achieving the objectives, different suggestions for delivery may of course be made. Based on the literature (Pedrinaci et al., 2013; King, 2012) and also in our experiences, 8 ideas are here suggested: the earth is a complex system in which the rocks, water, air, and life constantly interact; the origin of the earth is close to that of the solar system and its history is recorded in the rocks; water and air make the earth a special planet; life evolves and interacts with the earth modifying each other; plate tectonics is an explanatory theory for the out part of the earth; mankind depends of the planet for resources but must do so in a sustainable way; there are geological hazards relevant and danger for mankind; scientists seek explanations for the way the earth works through repeatable observations and interpretations.

Physics

Research has shown that students in higher education have conceptual difficulties in understanding physics concepts, namely in introductory physics courses (Costa, Oliveira, & Oliveira, 2012). Research conducted by the 2nd author of this chapter showed that a considerable number of futures physics and chemistry teachers still present some alternative conceptions about basic physics and chemistry concepts when they attend specific didactics classes, almost at the end of their formation (Costa, 2000). One way we suggest to change this situation is to use a didactical approach in physics classes, namely lectures and in introductory courses where fundamental concepts are taught, which challenges students' prior ideas. This didactical approach, which is being used by some researchers and teachers with success (Oliveira and Oliveira, 2013) has also the advantage of making the student more active in the lectures. The conceptual framework underlying the didactical approach is based on the instructional model developed since the late 1990's by Eric Mazur. Its landmark is the conceptual question, that is a question which usually does not include mathematical calculations but is focused on basic physic concepts. The didactic sequence in a lecture includes the main following steps: (a) the lecture is divided into a set of short presentations by the teacher (around 15 min), each focused on a central physics concept/theory; (b) after each short presentation, students are given a conceptual question, with a set of options with possible replies, in order to allow some time to verify how the students understanding of the physics concept/theory students occurred (e.g., of a conceptual questions: what happens to a bright and sonorous alarm clock when we put it in vacuum; with four options); (c) students are given a few minutes to answer individually and then they are asked to explain their answer to the other students sitting around them. The teacher encourages students to provide the reasoning behind their answers; (d) the students give their answers to the class; and (e) according to the students' answers the teacher decides what to do next (e.g., moves to the next topic if the majority of the answers are correct).

Complementing, in some way, what has been said, the authors argue that science teachers education in general have to take into account for the near future a quite relevant issue. From the beginning of the 21st century, a new scientific domain—science for sustainability—began to be developed (Vilches & Gil-Perez, 2015). The aim is to address the current situation of planetary emergency, and therefore, to help promote the transition to a sustainable world. It is another way of doing science in a new period of the history of our planet—the Anthropocene—in which large and disastrous changes, mainly due to the human beings action, are taking place threatening the survival of our own human species.

General Pedagogical Knowledge (PK)

For the sake of brevity, we will not refer to all aspects of the curricular units and domains belonging to general PK but rather refer to their main strengths and areas for improvements illustrating our discourse with some examples from the preparation of science teacher from our university.

New challenges to science teacher education implied the need to include new aims, contents, and even perspectives in general PK. Some examples, introduced in our university after the BP, are indicated below:

- "To understand and be able to reflect and discuss about school failure, violence, and indiscipline in schools and school–family relationships": new aim in sociology of education;
- "Development and learning in the future: e-learning, teacher and student roles, NTIC, psycho-pedagogical effects": new topics introduced in Psychology of Education;
- "From a national to a European teacher": new transversal perspective in science teacher education; and
- From a teacher with a "disciplinary and a canonic view" to a teacher with a "sustained view of science" and therefore with obvious implication with society.

However, the BP has diminished the hours (and ECTS) for the disciplines in the area of educational sciences, and therefore, some including general PK couldn't be included any longer in the formation path of a science teacher. As referred to before, there is not also space for a discipline in special needs education in our university, which may also be an aspect to improve.

A last point refers to the area of assessment and evaluation, considering an important issue namely to improve the Portuguese school system in Portugal (OECD, n.d.); therefore, a topic which should also have a relevant position in science teachers preparation. However, this so far doesn't happen in Portugal, namely because there is no space in the curricula. For example, at the University of Aveiro, there is an optional discipline for future biology/geology and physics/chemistry teachers called Assessment/Evaluation and Quality in Education with 6 ECTS. Indeed, several recent reports conducted by OECD about the educational system in Portugal refer to the inclusion of the topic of evaluation and assessment in teacher preparation, as assessment and evaluation can be seen as a way to improve the quality of education.

Pedagogical Content Knowledge (PCK)

Concerning PCK, the main general strengths identified by the authors of this chapter in Portugal are: The strong articulation between the topics approached in specific didactics and the ones emerged from the research in didactics of the disciplines to be taught; the way science teaching practice is organized, namely, the partnership with the schools. Notice the investment that many universities have made in developing specialized courses, namely academic masters, for school supervisors; the inclusion of a perspective not only of a "reflective practitioner" but also a "teacher as a researcher."

In specific terms we will refer now to a topic—practical work (PW) in science teaching—which is one where a strong investment has been made in teacher education in Portugal, namely by higher institutions (in teachers preparation), research, and by the government (namely, in the program "Ciência Viva" which started in the 90's of the 20th century but is still a landmark in Portugal (Gago, 1990). This topic also illustrates the strong ability in our country to articulate science education in several contexts (formal, non-formal, and informal). Also, PW means curricular and extracurricular experiences in a variety of settings (i.e., different environments) at school or outdoor—field, science centers, museums, urban spaces—implying that the student is an active subject in his/her own learning process, interacting with materials to observe and understand the natural world (Marques & Praia 2009; Bonito, 2012). Laboratory work (Tobin, 1990), experimental work (Leite, 2001), or field work (Orion, 1993) are also included within this terminology. Fieldwork activities, which, particularly in biology and geology teaching, play a crucial role (Rebelo, Marques, & Costa, 2011). Research shows that students have favorable attitude to this type of activities, and the use of appropriate strategies contribute to better learning. Regarding improvements and challenges to science teacher education in Portugal, the authors consulted three Portuguese senior researchers and science teacher educators.[1] Complementing their writings with what was written thus far, we listed below some improvements and challenges to science teacher's education in Portugal: The relevance of the students' development of a holistic picture of science through broad, interdisciplinary and balanced teaching and learning strategies, bearing in mind that science is also culture (Toharia, 2004); the importance of bearing in mind the most current and up to date scientific data about what is required for a learner to reach deep understanding and, therefore, to plan for effective teaching (Bransford, Brown, & Cocking, 1999); and the recognition that more and better practical work in the science curriculum and, therefore, at the classroom is needed. Learning is very often more effective when it incorporates hands-on activities. Practical work makes phenomena more real, arouses and maintains interest and promotes a logical and scientific reasoning (Science Community Representing Education, 2008). From the authors' view, this is particularly relevant when one argues that the real world interaction, the way and how it works, and the mankind ethical behavior are crucial issues, taking into account the main concerns of the science teaching and learning process. Taking on board a "neuroeducation" perspective: assuming that "teachers are more than anyone else, the key for educational change" (Hargreeves, 1998, p. 8), educators should be keen to develop teacher education based on updated scientific data; for instance, those related to the way the brain influences learning. In articulation with the cognitive psychology, the knowledge emerging from the neurosciences research gives important contributions for a better understanding of the teaching and learning process and enables the designing of new strategies to help both those who learn and those who teach. Therefore, Mora (2014) believes that "neuroeducation is a new vision of education based on the brain." This means that data from this new research area is useful to increase knowledge on how the brain works in a psychological and sociological context. The main aim, as far as teacher education is concerned, is to improve the students' underlying learning processes and, thus, to provide better teaching methods for teachers.

Ortiz (2013), for instance, discloses that brain research reveals that the following aspects are desirable for a good learning achievement: (a) a good emotional adaptability, (b) a personal effort towards a lifelong learning attitude, (c) a greater focus on problem solving, and (d) teaching organized on the basis of the brain development rather than on educational programs.

FINAL COMMENTS

The following important aspects should be mentioned:

- The preparation of science secondary school teachers—biology/ geology and physics/ chemistry—is made by two different master's courses that have some themes in common (i.e., psychology, sociology, methodology, . . .) and with others very specific (i.e., didactic of the teaching area, . . .).
- The definition of four scientific areas in science teacher education, enabling the achievement of a broad and balanced view about teaching in general and science teaching in particular, is in accordance both with law and educational research suggestions. A set of specific curricular units is assigned to each scientific area and it varies from institution to institution. The weight of each area is indicated by the amount of allocated ECTS, for instance the specific didactic and the introduction to professional practice areas reaches 60%.
- The initial teaching practice is an opportunity for a better understanding of the nature of the educational experience—a complex and problem-based issue—carried out in an ethical and social context.
- The design, development, and conclusion of the research project, rooted on the student's school experience and, therefore, linking teaching activities and educational research.
- Quality assurance requirements of teacher education courses are checked by an independent agency that can require the improvement of each course or even determine its non-approval.
- School Management . . .

As far as the future for the teacher education is concerned, the authors start from Shulman's "what do teachers need to know?" (Shulman, 1986), working as a popular avenue for quality teacher preparation. Indeed many other authors (for example, Ball, Thames and Phelps, 2008) refer to Shulman's work as a landmark in the conceptualization of teachers' knowledge and the advances it has produced in the field. Five aspects will be here underlined to promote focus on the classroom in particular and on the school in general:

- *Teaching area background:* No doubt about the importance of teachers mastering the scientific content, meanwhile its articulation with world issues is crucial, framing them in a multidisciplinary context; to become familiar both with the nature of the scientific knowledge, and with the processes of the scientific inquiry is also relevant, as well as with its limits and validity. The aim is to develop science teaching toward informing students about the natural systems and help the learners to achieve enough evidence-based knowledge to make scientific supported personal judgments to lead more ethic, cultural, healthy, comfortable, and environmentally sustainable lives. The perspective of the new scientific domain—science for sustainability—should be here underlined;

- *Pedagogical performance:* The designing and implementation of science teaching plans, promoting satisfactory students' understanding of the world, requires a demanding teacher conceptual framework crossing knowledge from different areas such as, psychology and sociology of education, curricular development, didactics, education technology, or evaluation. It is highly desirable that teachers understand the importance of teaching methodologies focused, for instance, on inquiry-based learning. Teachers must also recognize the relevance of, either a reflexive attitude based on their own teaching practices perceptions, or the discussion between peers. Science teaching requires that attention might be given to practical work, approached in an investigative way within a problem based learning perspective.
- *Dialogue educational research-teaching practice:* The authors assume the importance of educational research recommendations and also the difficult dialogue with the teaching practice; on that basis, all the attempts to promote this dialogue are positive. Investing in learning communities, taking advantage of ICT, can contribute to enrichment of the discussion and sharing between science teachers and science educators. Therefore, the educational research dimension in teacher education programs must be strengthened, either in initial or on-going professional learning towards a better quality of science teaching.
- *Emerging guidelines from new research fields:* Initial and in-service teacher education, taking into account its complexity and assuming that teachers are the key for the educational change needed, must integrate update contributions, coming from different areas of knowledge. For instance, cognitive neurosciences and cognitive psychology aim to understand better both the nature of learning and the nature of teaching, providing a new vision of science teaching, and therefore, their contribution cannot be neglected. Educational strategies should integrate what is currently known about the brain mechanisms underlying the learning process.
- *Ongoing teachers professional learning:* Assuming that there is demand and consequent need for more update science education at schools, so there is also a demand for many more qualified science teach-

ers. Important issues concerned with the challenge of conditions of service and also the organizational habits in several schools that do not facilitate the regular professional dialogue between teachers, political decision makers, and science educational researchers. This dialogue aims a science teachers' preparation towards a teaching approach that enables students' achievement of

- a healthy, safe, and responsible environmental attitude;
- a critical perspective concerned with the reliability of information provided by the media; and
- a holistic view about our place in the universe, recognizing scientific limitations and the potential of an adequate complement with other fields such as philosophy and religions.

NOTE

1. The authors acknowledge the kind contributions by colleagues Professor Cecilia Galvão, Professor Vitor Trindade, and Dr. Jorge Bonito.

REFERENCES

About A3ES. (n.d.). Retrieved from http://www.a3es.pt/en/about-a3es

Alarcão, I., Freitas, C. V., Ponte, J. P., Alarcão, J., & Tavares, M. J. F. (1997). *A formação de professores no Portugal de hoje* [Teacher Education in Portugal Nowadays]. Consultado em 25 maio 2015 em. Retrieved from http://www.educ.fc.ul.pt/docentes/jponte/docs-pt/97-Alarcao (CRUP).

Amaral, A., Tavares, O., & Cardoso, S. (2011, November). Regaining Trust. Is it possible? Paper presented at the *VI European Quality Assurance Forum*. Antwerp. Retrieved from http://www.a3es.pt/sites/default/files/30.%20Regaining%20Trust.pdf

Ball, D., Thames, M., & Phelps, G. (2008). Content knowledge for teaching. What makes it special. *Journal of Teacher Education, 59*(5), 389–407.

Bonito, J. (2012). *Panoramas Atuais Acerca Do Ensino Das Ciências* [Current perspectives about science teaching]. Boa Vista, Cape Verde: UFRR.

Bozzeli, F. C., & Nardi, R. (2012). Interações e o uso de analogias no ensino da física [Interactions and use of analogies in the physics teaching]. Investigações em *Ensino das Ciências, 17*(1), 81–107.

Bransford, J., Brown, A., & Cocking, R. (1999). *How people learn: Brain, mind experience, and school.* Washington, DC: National Academy Press.

Campos, B. P. (1995). *Formação de professores em Portugal* [Teacher Education in Portugal]. Lisbon, Portugal: Instituto de Inovação Educacional.

Carvalho, A., & Gil-pérez, D. (2011). *Formação de Professores de Ciências* [Science teachers education]. 10. São Paulo, Brazil: Cortez.

Costa, N. (2000). New Challenges to Initial Teacher Training: how the University of Aveiro (Portugal) is responding to change? *Actas da Conferência Internacional* [Proceedings of the International Conference] *Education and Challenges of New Millennium* (pp. 8–11). Plovdiv, Bulgaria: University of Plovdiv.

Costa, N. (2015). Challenging physics lessons through questioning and collaborative work: Research developed in Portuguese Higher Education Institutions. Extended Summary in H. Pedrosa-de-Jesus and M. Watts (Eds.), *Seminar Proceedings of the 3rd International Seminar on Research on Questioning "Critical inquiry and academic growth in Higher Education"* (pp. 49–55). Aveiro, Portugal: University of Aveiro. Retrieved from http://edaun.web.ua.pt/wp-content/uploads/2015/02/Seminar-proceedings.pdf

Costa, N., Graça, B., & Marques, L. (2003). Bridging the gap between science education research and practices: A study based on academics'opinions. *Procedures of the Meeting Teaching and Learning in Higher Education: New Trends and Innovations* [CD version]. Aveiro, Portugal: University of Aveiro.

Costa, N., Oliveira, P. E., & Oliveira, C. (2012). O Ensino da Física em Cursos de Engenharia: Elementos Potenciadores do Sucesso dos Estudantes [The teaching of physics in engineering courses: elements that enhance students success]. In J. E. Fuentes Betancour, A. P. Perdomo, O. A. Calzadilla Amaya (Eds.), *Proceedings of Taller Iberoamericano de la Enseñaza da la Física Universitaria* [CD version]. Havana, Cuba.

Cruz, E., Pombo, L., & Costa, N. (2008). Dez anos (1997–2007) de estudos sobre o impacto de Cursos de Mestrado nas práticas de Professores de Ciências em Portugal [Ten years (1997–2007) of studies on the impact of Masters Courses in the practices of Science teachers in Portugal]. *Revista Brasileira de Pesquisa em Educação em Ciências,* 8(1). 1–22.

European Commission. (n.d.). Common European Principles for Teacher Competences and Qualifications. Retrieved from http://www.pef.uni-lj.si/bologna/dokumenti/eu-common-principles.pdf

Estrela, M. T. (2014). Velhas e novas profissionalidades, velhos e novos profissionalismos: Tensões, paradoxos, progressos e retrocessos [Old and new professionals, old and new professionalism: tensions, paradoxes, progresses and setbacks]. *Investigar em Educação, 2*(2), 5–30.

Florian, L., & Pantic, N. (2013). *Learning to teach: Part 2: Exploring the distinctive contribution of higher education to teacher education.* York, England: The Higher Education Press.

Furlong, J. (2013).The policy context: The changing nature of initial teacher education. *Research Intelligence, 120,* 9–10.

Gago, J. M. (1990). *Manifesto para a Ciência em Portugal* [Manifesto for Science in Portugal]. Lisbon, Portugal: Gradiva.

Galvão, C., Reis, P., Freire, A., & Oliveira, T. (2007). Science curriculum in Portugal: From the development to the evaluation of students' competences. In D. Waddington, P. Nentwig, S. Schanze. (Eds.), *Making it comparable: Standards in science education* (pp. 237–253). Munster, Germany: Wasmann.

Hanna, S., & Lisa, G. (2008). Analysing how scientists explain their research: A rubric for measuring the effectiveness of scientific explanations. *International Journal of Science Education, 30*(11), 1441–1467.

Hargreaves, A. (1998). *Os professores em tempos de mudança: trabalho e cultura dos professores na idade pós-moderna* [Teachers in times of change: Work and culture of teachers in the postmodern age]. Lisbon, Portugal: McGraw-Hill.

King, C. (2012). Earth-science. *School Science Review, 94*(347), 23–24.

Leite, L. (2001). Contributos para uma Utilização mais Fundamentada do Trabalho Laboratorial no Ensino das Ciências [Contributions to a more sustained use of laboratory work in science teaching]. In H. V. Caetano & M. G. Santos (Eds.), *Cadernos Didáticos de Ciências 1* (pp. 79–97). Lisbon, Portugal: Departamento do Ensino Secundário.

Lima, L.C., Azevedo, M.L.N., & Catani, A.M. (2008). O Processo de Bolonha, a Avaliação da Educação Superior e Algumas Considerações sobre Universidade Nova [The Bologna process, the assessment of higher education and few comments about the new university]. *Avaliação, 13*(1), 7–36.

Loughran, J., Berry, A., & Mulhall, P. (2012). *Understanding and developing science teachers' pedagogical content knowledge.* Rotterdam, the Netherlands: Sense.

Marques, L., & Praia, J. (2009). Educação em Ciência: Atividades exteriores à sala de aula [Science education: Outdoor activities]. *Terrae Didática, 5*(1), 10–26.

Marques, M. M., Loureiro, M. J., & Marques, L. (2015). The dynamics of an online community of practice involving teachers and researchers. *Professional Development in Education,* 42(2), 1–23. Retrieved from http://www.tandfonline.com/doi/abs/10.1080/19415257.2014.997396#.VffR9dJViko)

Mora, F. (2014). *Neuroeducación. Sólo se puede aprender aquello que se ama* [Neuroeducation. You can only learn what you love]. Madrid, Spain: Alianza Editorial.

Nascimento, A. & Vaz-Rebelo, P. (2004). A formação psicopedagógica nos ramos de formação educacional da FCTUC: balanço e perspetivas [Psychopedagogical training in the educational fields of the FCTUC: balance and perspectives]. *Actas do III Simpósio Ensinodas Ciências e da Matemática: formação de professores, perspetivas e desafios* (p. 125). Lisbon, Portugal: Departamento de Educação da Faculdade de Ciências da Universidade de Lisboa.

Organization of Economic Co-operation on Development. (2014). *Education policy outlook: Portugal.* Retrieved from www.oecd.org/edu/policyoutlook.htm

Oldroyd, D. (1996). *Thinking about the Earth: A history of ideas in geology.* London, England: The Athlone Press.

Oliveira, P., & Oliveira, C. (2013). Using conceptual questions to promote motivation and learning in physics lectures. *European Journal of Engineering Education, 38*(4), 417–427.

Orion, N. (1993). A model for the development and implementation of field trips as an integral part of science curriculum. *School Science & Mathematics, 93*(6), 325–331.

Ortiz, T. (2013). *Neurociencia y educación* [Neuroscience and education]. Madrid, Spain: Alianza Editorial.

Osborne, J., & Dilon, J. (2008). Science education in Europe: Critical reflections. *A Report to the Nuffield Foundation.* London, England: King's College London.

Pedrinaci, E., Alcalde, S., Alfaro, P., Almodóvar, G. et al. (2013). Alfabetización en Ciencias de la Tierra [Literacy in Earth Sciences]. *Enseñanza de las Ciencias de la Tierra, 21*(2), 117–129.

Ponte, J. P. (2006). Os desafios do Processo de Bolonha para a formação inicial de professores [The challenges of the Bologna Process for initial teacher training]. *Revista da Educação, 14*(1), 19–36.

Rebelo, D., Marques, L., & Costa, N. (2011). Actividades en ambientes exteriores al aula en la Educación en Ciencias: Contribuciones par su operatividad [Outdoor activities in science education: Contributions for implementation]. *Enseñanza de las Ciencias de la Tierra, 19*(1), 15–26.

Russell, T. (1993). Critical attributes of a reflective teacher: Is agreement possible? In J. Calderhead & P. Gates (Eds.), *Conceptualizing reflection in teacher development* (pp.144–154). London, England: The Falmer Press.

Roldão, C. (2007). Função docente: natureza e constância do conhecimento profissional [Teaching role: Nature and persistence of professional knowledge]. *Revista Brasileira de Educação, 2*(34),94–103.

Sá-Chaves, I. (2005). *Os "Portfolios" Reflexivos (Também) Traz, Gente Dentro. Reflexões em torno do seu uso na humanização dos processo educativos* [Reflexives portfolios also bring people inside. Reflections about their role for the humanization of educational processes]. Porto. Porto Editora. Coleção Cidine.

Schon, D. (1983). *The reflective practitioner: How professionals think in action.* New York, NY: Basic Books.

Science Community Representing Education. (2008). *Practical work in science: A report and proposal for a strategic framework.* London, England: Science Community Representing Education.

Shulman, L. (1986). Those who understand: Knowledge growth in teaching. *Educational Researcher, 15*, 4–14.

Tobin, K. G. (1990), Research on science laboratory activities: In pursuit of better questions and answers to improve learning. *School Science and Mathematics, 90*, 403–418.

Toharia, M. (2004). La Ciencia también es Cultura [Science is also culture]. *Enseñanza de las Ciencias de la Tierra*, 12, 20–23.

Valente, M. O. (2002). História da formação de professores na Faculdade de Ciências de Lisboa e do Departamento de Educação [History of teacher education at the Lisbon Faculty of Sciences and Education Department]. *Revista de Educação, 11*(1), 7–15.

Vieira, F., Silva, J. L., Vilaça, T., Parente, C., Vieira, F., Almeida, M. J., . . . & Silva, A. (2013). O papel da Investigação na Prática Pedagógica dos Mestrados em Ensino [The role of research in teaching masters courses' school practice]. In *Atas do XII Congresso Internacional Galego-Português de Psicopedagogia* (pp. 2641–2655). Braga, Portugal: Universidade do Minho.

Vilches, A., & Gil-Perez, D. (2015). Ciencia de la Sostenibilidade: Una nueva disciplina o un nuevo enfoque para todas las disciplinas? [Sustainability Science: A new discipline or a new approach for all disciplines?] *Revista Iberoamericana de Educación, 16*(1), 39–60.

CHAPTER 6

SCIENCE TEACHER PREPARATION IN TURKEY

Hakan Türkmen
Ege University

The Republic of Turkey was founded on the ruins of Ottoman Empire in 1923 by Mustafa Kemal Ataturk. Turkey adopted wide-ranging social and political reforms under his authoritarian leadership to take her place among the "developed nations" of the world. Turkey is a bridge between Europe and Asia with a population of 78 million (2015 census) and the only country among Islamic countries which has declared secularism.

The Turkish education system (see Figure 6.1) includes *preschool* (ages 3–5.5); *primary* (ages 5.5–10); *lower secondary* (middle school, ages 10–14) dividing into 3 types: general lower secondary school, Anatolian school, and Imam Hatip school; *upper secondary* (high school, ages 14–18) dividing into 6 types: general upper secondary school (called Lise), secondary vocational school (called Meslek lisesi), general upper secondary and vocational school (called Teknik lise), science school (called Fen Lisesi), Anatolian high school, and Imam Hatip high school; and *higher education* and *non-formal education* including all the activities organized outside or alongside the school (such as cram schools and private lessons).

Model Science Teacher Preparation Programs, pages 129–149

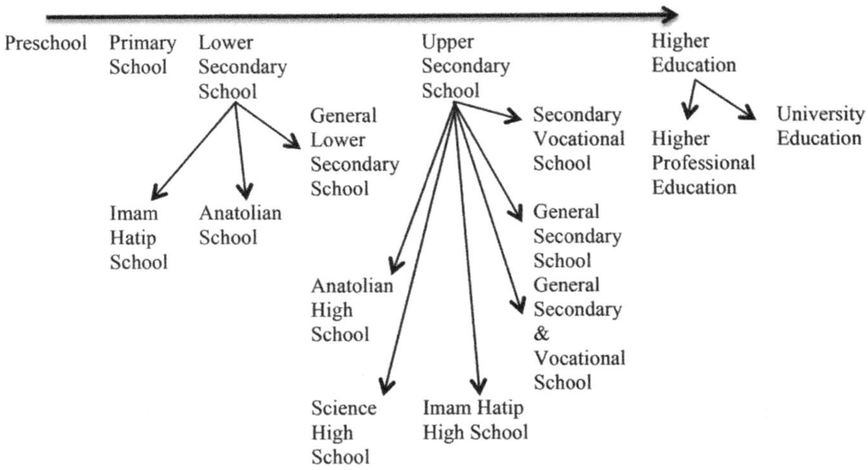

Figure 6.1 Turkish education system.

Formal education statistics reported in Ministry of National Education (MoNE) in 2014–15 that there are 985,013 preschool students in 26,972 preschools; 5,230,878 primary students in 27,544 primary schools; 5,069,683 lower secondary school students in 16,969 lower secondary schools; 583,344 students in private schools; and 5,691,071 upper secondary school students in public and private upper secondary schools. And 919,393 teachers are working in 59,509 schools in the 2014–15 academic years. But, the number of schools is not adequate when compared to the number of children of school age. Classes average 27 children in the primary schools, 28 in the lower secondary schools, and teacher–student ratios are 1:18 for primary schools, 1:17 for lower secondary schools, and 1:14 for upper secondary schools (MoNE, 2015).

Every Turkish citizen has a right to education which is free of charge for compulsory primary and secondary education. This system was centralized by the Law of Unification of Instruction Act in 1924. The MoNE has the responsibility to run educational administration all over the country. The MoNE must draw up curricula which must be used in all schools; coordinate the work of official, private, and voluntary organizations; design and build schools; develop educational materials and so on. By the acceptance of Basic Law of National Education in 1973, teachers should be educated in higher education institutions (universities), and the teacher education responsibility of MoNE was transferred to the Higher Education Council (HEC) in 1981 (Akyuz, 2004; Çakıroglu & Çakıroglu, 2003; Deniz & Sahin, 2006; Tarman, 2010; Turkmen, 2007; Yuksel, 2008). HEC has the authority for the regulation of higher education. And although it operates as a fully independent national institution without any political or government affiliation, in reality every different political aspect ignored this policy and wanted to have control HEC. This is obvious that every political party declared when we selected as a government in the national selection, one of our duty change HEC structure or closed HEC in order to give more freedom to universities. The HEC determines the requirements for the promotion of academicians from lecturer to professor and, even decides what academician is qualified to promote to associate professor by associate professor oral exam. Moreover the HEC determine the standards for university degrees and, defines the structure of teacher education programs, such as what education program is opened or not, the length of each field of programs, the number of credits, the name, and content of courses (Grossman, Sands, & Brittingham, 2010; Ozden 2007).

Until the founding of HEC, three-year institutes of education had taken the responsibility of training teachers for secondary schools, and the length of teacher programs in institutions of education was increased from three years to four years in 1976. After 1981, this program was reorganized under the new name as Higher Teacher Training Colleges with new departments such as physics, Chemistry, etc. In 1982, they were connected to faculties of education in the universities (HEC, 1999). Therefore, the HEC has restructured and continues to develop and modify all the teacher education programs as necessary along with European Union (EU) requirements and modern educational philosophy. One of the important necessities, which are standards and accreditation of beginning from subject area to departments in universities and universities, is not succeed yet (Guven, 2008; Grossman et al., 2010).

CURRICULUM DEVELOPMENT STUDIES

Curriculum development studies in Turkey began in the 1920s with the recommendations of the American philosopher, John Dewey (Gokmenoglu & Eret, 2011; Unal & Unal, 2010). Since 1924, Turkey has always valued education and made efforts to establish a modern education system. So many science education programs were developed and applied by MoNE. Many recent reforms have been supported by international organizations, in certain cases beginning as pilot projects designed to transform national education policy. One of the recent efforts was the National Education Development Project that featured two major components between the years 1991 and 1999. The first was a reform of primary and secondary education school project managed by MoNE in terms of infrastructure, textbooks, and educational management. The second, managed by HEC in late 1994, was a pre-service teacher education project. This project aimed for every different education program to have the same curriculum nationwide, for example the science teacher education program should be the same and used in all Turkish education faculties. It was called the package program. With the new curriculum and an increased emphasis on teaching models, pre-service teachers spent much more time practicing in schools than in the older model, which emphasized subject contents in order to enhance the professional development of pre-service teachers (Akyuz, 2004; Çakıroglu & Çakıroglu, 2003; Deniz & Sahin, 2006; Tarman, 2010; Turkmen, 2007; Yuksel, 2008). Moreover the HEC, in the light of the pre-service teacher education project, restructured faculties of education and prepared an accreditation model for them.

Another effort was The Basic Education Program in 1997, and then the Secondary Project in 2006, sponsored by the World Band, and aimed to improve the quality of education at these different levels. Compulsory education was just 5 years (primary school) by 1997, and it was extended from 5 (just graduating from primary schools) to 8 years (just graduating from secondary schools). The Turkish education system has been completely reorganized by this effort and the primary and lower secondary science curriculum was revised by the MoNE with the Secondary Project. Because of these reform efforts, many Turkish teachers have been sent to the United States, France, and the United Kingdom via the National Educational Project supported by the World Bank since 1993. More specifically, many scientists have been sent to study for master and PhD degrees in science education to the United States, France, and the United Kingdom since 1997 (Turkmen, 2005). Also, around the same time, many Turkish universities began to open science education departments or programs in their universities to address this new focus. Today, there are 193 universities and 97 faculty of education in Turkey. Currently, there are 70 science education departments or programs in Turkey, and the number of science education programs or departments continues to increase. While two of the public universities are more than 100 years old, according to the rankings of an education counseling firm in the United Kingdom in 2014, three Turkish universities are identified in the top 500 (MoNE, 2015).

In the meantime, as many research projects like Altan's study in 1998 indicate, teacher candidates have been educated as specialists in their subject areas, and pedagogy has been neglected. This was because many academicians in science education departments were well qualified in their own subject areas, such as science and research methods, but they were not good at pedagogical aspects of education. These academicians were graduated from arts and science faculties. This was a serious issue in secondary science teacher education (Kavcar, 2002). As a result, HEC and MoNE co-operated and restructured teacher education programs. The revised program began in the 1998–99 academic year. After Turkish education faculties began to adapt the teaching and preparation of science programs, teacher candidates at universities began to be more involved with school experiences and teaching practice activities (hands-on activities). However, this effort was not seen as enough by researchers and education philosophers (Gokmenoglu & Eret, 2011; Guven, 2008; Koc, Isiksal, & Bulut, 2007; Ozden, 2007). Besides, results of international studies were somewhat alarming and provided impetus for urgent action.

Many researchers declared that the reasons for these results are related to the traditional education system, inadequate educational standards, inadequate preparation of teachers, and limited budgets for education. Thus, the existing infrastructure, the limited or inappropriate resources, inadequate university teachers and educational practices did not foster the development of Turkish education system. Many researchers reached the consensus that the Turkish education system needed to initiate big revisions in the light of modern educational perspectives (Bilgen, 1999; Elmas, Öztürk, Irmak, & Cobern, 2014; Grossman et al., 2010; Ozden, 2007; Tarman, 2010; Turkmen, 2005; Ustuner, 2004). Indeed, many teacher candidates entered university with an insufficient understanding of school science and university teachers were usually using lecture-based methods that emphasize rote learning of disconnected facts (Ozden, 2007; Ustuner, 2004; Yuksel, 2008). As a result of that fact, teacher candidates' understanding of science was poor and their pedagogical practices were inadequate. When these teacher candidates become teachers at schools, they will not teach enough science to the next generation of students.

The MoNE started to examine and evaluate several modern countries' teacher education programs and school curriculums, such as those of the United States, Britain, Germany, and Finland. In light of these examinations, the teacher education programs and school curriculums have been reconsidered again in Turkey, especially in the areas of learning theories, curriculum development, and educational technology. The U.S. national as well as state standards, benchmarks, and Project 2061 (Project 2061 is a long term research and development project and began its work in 1985. The main initiative focus of this project is to improve science education in the United States helping all Americans become literate in science, mathematics, and technology); and EU Educational policies, such as the Bologna Process; European Credit Transfer and Accumulation System; Erasmus program; and Socrates I–II programs have all gained wide acceptance (Akyuz, 2004; Bilgen, 1999; Cakıroglu & Cakıroglu, 2003; Elmas et al., 2014; Grossman et al., 2010; Guven, 2008; Koc et al., 2007; Turkmen, 2005; Turkmen, 2007).

Scientists and educators worried that students were not being prepared well enough to live in tomorrow's technology and science-oriented world. Following an examination of modern countries' programs, in 2004, school curriculums were modified again (Deniz & Sahin, 2006; Tarman, 2010;

Turkmen, 2005; Turkmen, 2007; Yuksel, 2008), first at the primary level and then at lower and upper secondary.

After a pilot study was applied in 9 providences and 120 schools, renewed curriculum has been implemented since 2005 in all of the primary schools. This reform has been described as "constructivist education reform." With constructivist education reform, the purpose of the science curriculum was widened, but content was diminished, to include preparing students to be scientifically and technologically literate citizens who are able to use scientific facts and technology in their lives. This reform expects students will have skills including critical and reflective thinking, creativity, communication, problem solving, and investigation, and emphasize the driven decision-making process, and the use of information technologies (Aksit, 2007; MoNE, 2006). The new curriculum reform brought major changes in philosophy of instruction, teaching styles, teacher and student roles, and curriculum organization based on a constructivist approach to instruction. The curriculum emphasized student-centered strategies, scientific skills, and high-level order thinking process. Within the framework of a spiral curriculum, topics were expanded and elaborated throughout the years. A thematic approach was used in the organization of the content, and there are four learning areas: *living organisms and life, matter and change, physical events,* and *the earth and the universe.* There were also three learning areas related to skills, attitude, and values, which are science process skills, science-technology-society-environment, and attitudes and values (MoNE, 2004).

The 2004 curriculum incorporated crucial changes about student and teacher roles. Student roles changed from passive listeners to active participants who investigate, question, and solve the problems on their own. The teacher is the "facilitator" who organizes the teaching environment, guides the learners during the activities, involves students in the decision-making process, encourages students to share and discuss their ideas, and makes connections between daily life examples and scientific concepts (Aksit, 2007; Deniz & Sahin, 2006; Grossman, Onkol, & Sands, 2007; Tarman, 2010; Turkmen, 2005; Turkmen, 2007; Yuksel, 2008). Another effect of the 2004 curriculum was about assessments. The aim of the assessment was to assess not only the end product but also the entire process with the aid of performance tasks, concept maps, structured grid, projects, and poster presentations (Grossman et al., 2007; Irez, 2006; Ozden, 2007). These changes in the new primary science curriculum had also some implications for the secondary science education in Turkey.

In 2011, secondary science education curriculum namely, biology curriculum (MoNE, 2011a), physics curriculum (MoNE, 2011b), and chemistry curriculum (MoNE, 2011c), changed considerably. The new upper secondary science (physics, biology, chemistry) curriculum was prepared within the same points of primary science curriculum, such as

TABLE 6.1 A Comparison of the Old Versus the New Curriculum

Old Curriculum	New Curriculum
Information does not change	Information changes
Education is for knowing	Education is for understanding
Teacher as the information provider	Teacher as the facilitator
Teacher as the only decision maker	Teacher and students make decisions
One-way communication	Two-way communication
Product-based	Process-based
School for individual's learning	School for everyone's learning
Parents do not know about education	Parent involvement is essential
Competency-based learning	Community-based learning
Norm-referenced assessment	Criterion-based assessment
Teacher knows the answers	There is more than one solution and the teacher may not know all the answers

Source: Koc, Isiksal, Bulut, 2007.

constructivist teaching approach, key concepts, spiral curriculum, and science process skills. In addition, some of the subdimensions of the learning areas were changed. For instance, in the science-technology-society-environment learning area, socio-scientific issues and sustainable development concepts were added. There was also a decrease in the total number of objectives throughout the curriculum. However, the major approach to science teaching, the topic structure, and the general aims of the curriculum, suggested teaching methods and spiral structure of the primary science curriculum stayed the same (see Table 6.1).

To sum up, the 2004 science and technology curriculum reform was comprehensive taking into account students, teachers, instructional materials, teaching approaches, and philosophies. The effectiveness of the 2004 curriculum was evaluated by Turkish academicians. For example, the Bozyılmaz and Bagci-Kilic (2005) study, the Erdogan (2005) study, and the Cakir (2005) study stated that scientific knowledge was diminished, scientific process skills increased, and science-technology-society connection was more focused on science topics than the last curriculum (cited in Turkmen 2007). Half of the program is theory and technique, a quarter of it is professional teaching theory and technique, and the rest is general cultural information. With new programs, teachers have more of an opportunity to study in depth their own branch. In the meantime, faculties of education have some flexibility to determine one third of the components of the program. That means they have the opportunity to increase their elective courses. Additionally, the ratio of general culture courses, like science, history, history of Turkish education, and an introduction to philosophy, was increased in the new program (MoNE, 2006).

With the Primary Education Law No. 6287 adopted on March 30, 2012, a radical decision was made in the Turkish education system and put into practice. This unexpected law is known by the public as 4 + 4 + 4. Through 4 + 4 + 4 projects, primary school length was decreased from 5 to 4 years, middle school was designed lower secondary and its length was increased from 3 to 4 years, and high school became upper secondary and its length was increased 3 to 4 years. This change was enacted suddenly and without any pilot implementation (Gun & Baskan, 2013). Moreover, compulsory education was also extended from 8 to 12 years and was defined as a "paradigm shift" by the MoNE" (MoNE, 2013). This rapid and sudden change with the 4 + 4 + 4 education system has brought along various problems and criticism. Many researchers and academicians state that current education efforts do not function completely (Gun & Baskan, 2013; Karadeniz, 2012; Memişoglu & Ismetoglu, 2013). Without any pilot implementation, without considering cost benefit analyses, without considering the opinions of academicians and civil associations of education, and without detailed academic studies, this effort is not based on a scientific approach and it has a possibility to harm education in Turkey.

Adding elective lessons, which are related to the field of religion, moral behavior, and values; foreign language; science and mathematics; art and sports; social science; and language and speech into 5th, 6th, 7th, 8th grades curriculum is a good movement, but if the government opens positions for teachers mostly in the field of religion, moral behavior, and values, it is a problem, because many science, art, technology, sports, and foreign language teachers have been waiting for available positions to be assigned in the MoNE. On the other hand, these elective lessons made total school hours in a week increased from 30 to 36/37 in public schools. This is another problem, because children's free time to be socialized in public and to spend time with family decreased.

Only increasing the number of Imam Hatip schools, Imam Hatip high schools, and Islamic religion schools, and not secondary vocational school, general secondary and vocational school (art and medical lab, etc.), and decreasing the school age from 7 to 5 or 5.5 may affect our children's future in a negative way (Gun & Baskan, 2013; Karadeniz, 2012; Memişoglu & Ismetoglu, 2013).

Besides applying new education programs, the Turkish government has sought assistance by sending teachers to western countries (especially the United States, England, Germany, and France) in order to study and learn about innovations in education and how these innovations may improve Turkish education. Unfortunately, these solutions have not resulted in progress on par with other modern countries up until this point in time.

Thus, within the current climate of Turkey, there is widespread dissatisfaction with the educational system. Government employees, teachers, and educators, as well as parents have all expressed this dissatisfaction (Organisation for Economic Co-operation and Development [OECD], 2013; Ozden, 2007; Tarman, 2010). These thoughts also have been supported by international exam results. For example, according to the Program for International Students Assessments (PISA) results, Turkey has made significant improvements in mathematics and science assessments, but remains below the OECD average in reading, mathematics, and science fields for the last 4 exams (PISA, 2003, 2006, 2009, 2012).

SCIENCE CURRICULUM IN LOWER SECONDARY SCHOOLS

The current lower secondary science curriculum focuses on 4 related areas (see Table 6.2). Learning areas consist of 4 main contents: *living organisms and life, matter and change, physical events,* and *the earth and the universe.* These areas are the same for all lower secondary levels and start in the 3rd grade of primary school. Within the framework of a spiral curriculum, topics are much more detailed throughout the years. For example, in the Earth and the Universe unit, students learn about earth movement, earth rotation and the day and year concept in the 4th grade; components of earth crust, effect of erosion and landslides in the earth crust, groundwater and surface water in the earth crust, air, earth, and water pollution in the 5th grade; dimensions and shapes of moon, earth, and sun, layers of earth, rotation of moon, the relationship of motion between earth and moon and results of that relationship, and the effect of the relationship of motion between earth and moon to daily life in the 6th grade; recognize sky objects, stars, constellations, solar system and planets, and conduct space researches in

TABLE 6.2 Science Teaching Programs Contents

Learning Areas (as Unit)	Skill	Affective Behaviors	Science-Technology-Society-Environment
• Living Organisms and Life • Matter and Change Physical Events • The Earth and Universe	• Scientific process skills • Life skills • Analytic thinking • Making decision • Creative thinking • Entrepreneurship • Communication • Team work	• Attitude • Motivation • Values • Responsibility	• Socio-scientific issues • Nature of science • Science Technology relationship • Contribution of science in society • Consciousness of Sustainable development • Consciousness of science and career

the 7th grade; earthquakes and related essential concepts, fault lines in Turkey, protections of earthquakes, weather events in the atmosphere, seasons, climate, and global warming in the 8th grade. The skills, affective behaviors, and science-technology-society-environment areas are integrated into learning scientific knowledge areas.

The science lesson, as a core course, is 4 hours in a week, and science applications, as an elective course, is 2 hours for each grade.

By the end of the lower secondary schools, students have to take a high school entrance exam (TEOG) for transition from lower to upper secondary education system. The students will be tested from 6 lessons: Turkish language, mathematics, science, revolution history and kemalism, religious culture and moral knowledge, and foreign language. The latter is taken in two examinations, the first of which is applied at the end of the fall term in two days, and second of which is applied at the end of the spring term in two days. Students answer 20 multiple-choice test questions from each of 6 lessons in the examinations. Year-end success points of students in 6th, 7th, and 8th grade classes along with scores taken as a result of TEOG are subject to a certain average and high school. Moreover a placement process is implemented using the points obtained.

BIOLOGY, PHYSICS, CHEMISTRY CURRICULUM IN THE UPPER SECONDARY

The biology, physics, and chemistry core courses are 2 hours in a week for 9th grade in all types of upper secondary schools, but they are elective courses in 10th, 11th, and 12th grades. As elective courses, in general upper secondary schools, there are 2(3) hours for the 10th and 12th grades, 2(4) hours for the 11th grades in a week; at the Anatolian High School, 3 hours for the 10th and 12th grades, 4 hours for the 11th grades in a week; in the science schools, 4 hours for 10th and 12th grades, 5 hours for 11th grades in a week. There are neither cores nor elective biology, physics, and chemistry courses in other types of upper secondary schools.

The biology curriculum includes biology in life, living organisms' world, and current environment problems topics in the 9th grade; reproduction, essential principles in genetic, and our earth topics in the 10th grade; energy conversion in living organisms, human physiology, science of behavior topics in the 11th grade; and from genes to protein, plant biology, community and populations of ecology, and beginning of life and evolution topics in the 12th grade.

The physics curriculum includes entrance of physics, matter and its features, motion and force, energy, heat and temperature topics in the 9th grade; pressure and buoyant force, electricity and magnetism, waves, and optics topics in the 10th grade; motion and force II, and electricity and magnetism II topics in the 11th grade; and regular circular motion, basic harmonic motion, waves mechanism, atom physics and radioactivity, modern physics, the application of modern physics in technology topics in the 12th grade.

The chemistry curriculum includes entrance of chemistry, atom and periodic system, chemical species interactions, and states of matter topics in the 9th grade; acid and bases, mixtures, energy in industry and living organisms, and chemistry in everywhere topics in the 10th grade; modern atom theory, chemical measurements, gases, liquids solutions, chemistry and energy, and chemical equilibrium and rate topics in the 11th grade; and chemistry and electricity, entrance of carbon chemistry, organic compounds, and chemistry in our life topics in the 12th grade.

In order to gain admission to an undergraduate program in a higher education institution, all upper secondary graduated students have to take university entrance examinations administered by the Student Selection and Placement Center (ÖSYM). This process consists of two stages: the higher education examination, called Yüksek Öğretime Geçis Sınavı (YGS) and the bachelor placement examination, called Lisans Yerleştirme Sınavı (LYS). The YGS exam takes place in April; the LYS exam takes place in June. The placement of the students who want to study in a 2-year undergraduate program is based on the scores from YGS, except for vocational high school graduates who apply for placement in 2-year vocational school programs, which are compatible with their high school majors and their grade point averages, without an entrance examination.

Moreover, students willing to go to the School of Physical Education and Sport, and the State Turkish Music Conservatory have to take YGS as a prerequisite for placement and an aptitude test as defined by the universities. As seen in Table 6.3, the YGS exam takes place in one session (160 min.) and has 4 fields of questions: Turkish language (40 questions), social science (40 questions), essential math (40 questions), and science (40 questions). LYS is applied in 5 separate sessions. Students have to answer 80 math and geometric questions in 135 minutes in the LYS-1 session; 90 physics, chemistry, and biology questions in 135 minutes in the LYS-2; 80 Turkish language and literature, and geology-I questions in 120 minutes in the LYS-3; 90 history, philosophy, and geology-II questions in 135 minutes in the LYS-4; 80 foreign language questions in 120 minutes in the LYS-5 (Öğrenci Seçme ve Yerleştirme Merkezi [OSYM], 2015).

SCIENCE TEACHER UNDERGRADUATE PROGRAMS

Undergraduate programs last for 4 years, and their common curriculum is designed by the HEC. Turkish science teacher candidates take 153 credits (132 hours lectures, 42 hours practicum courses) in four years.

TABLE 6.3 YGS and LYS Examinations

<table>
<tr><td>YGS</td><td>160 questions (q), 160 min.</td><td colspan="3">Turkish Language (40 q/40 min)</td><td colspan="3">Social Science (40 q/40min)</td><td colspan="3">Essential Math (40 q/40 min.)</td><td colspan="3">Science (40 q/40 min.)</td></tr>
<tr><td rowspan="5">LYS</td><td>LYS1: 80 questions, 135 min.</td><td colspan="12">Math (50 q/75 min.) and Geometric (30 q/60 min.)</td></tr>
<tr><td>LYS2: 90 questions, 135 min.</td><td colspan="4">Physics (30 q/45 min.)</td><td colspan="4">Chemistry (30 q/45 min.)</td><td colspan="4">Biology (30 q/45 min.)</td></tr>
<tr><td>LYS3: 80 questions, 120 min.</td><td colspan="8">Turkish Language and Literature (56 q/85 min.)</td><td colspan="4">Geology I (24 q/35 min.)</td></tr>
<tr><td>LYS4: 90 questions, 135 min.</td><td colspan="4">History (44 q/65 min.)</td><td colspan="4">Philosophy (32 q/50 min.)</td><td colspan="4">Geology II (14 q/20 min.)</td></tr>
<tr><td>LYS5: 80 questions, 120 min.</td><td colspan="12">Foreign language</td></tr>
</table>

Science content courses:

- general biology I & II (4 hours), general biology lab I & II (2 hours), human anatomy and physiology (2 hours), genetics and biotechnology (2 hours), evolution (2 hours), special topics in biology (2 hours)
- general chemistry I and II (4 hours), general chemistry Lab I and II (2 hours), analytic chemistry (2 hours), organic chemistry (2 hours)
- general physics I & II (4 hours) & III (2 hours), introduction modern physics (2 hours), general physics lab I & II (2 hours), special topics in physics (2 hours), astronomy (2 hours)
- general math I & II (4 hours), environmental science (3 hours), science and technology lab (2 hours), nature and history of science (2 hours), geology (2 hours)

Educational courses:

- introduction to educational science (3 hours), educational psychology (3 hours), principles and methods of instruction (3 hours), program development (3 hours), instructional technologies and designing teaching materials (3 hours), instructional methods in science education –II (4 hours), classroom management (2 hours), guidance (3 hours)

Science methods courses:

- statistics (2 hours), scientific research method (2 hours), measurement and evaluation (2 hours)

General courses (literature, cultural, computer, and foreign languages:

- principle of Atatürk and recent Turkish history I & II (2 hours), Turkish education eystem and echool management (2 hours), Turkish language I: writing expression (2 hours), Turkish language II: oral expression (2 hours), computer I & II (2 hours), foreign language I & II (2 hours), community service activities (3 hours)

Elective courses:

- 4 different courses (8 hours), offered by science education departments. These courses should be related science and general educational courses. These courses are available only 2nd year second semester or 4th year second semester for teachers candidates

Public School Practicum courses:

- school experience I (5 hours), teaching practice (8 hours) (HEC, 2005, 2009)

UPPER SECONDARY BIOLOGY, PHYSICS, AND CHEMISTRY UNDERGRADUATE PROGRAMS

Upper secondary biology, physics, and chemistry teachers' undergraduate programs in departments of faculties of education last for 5 years, and their common curriculum is designed by the HEC. But the standard 5 years education was decreased to 4 years by Turkish government in 2013. The program is very similar to a science teacher undergraduate program. The main difference is in science content courses. Each field (biology, physics, and chemistry) is specialized in their own field offering the deep science courses. The other educational and general literature, cultural, computer, and foreign languages courses are the same. As well as science teacher candidates in the lower secondary schools, upper secondary biology, physics, and chemistry teacher candidates practice with a "school experiences" course in the fall semester and a "teaching practice" course in the spring semester during their fourth year of education. Each semester, candidates are required to attend schools 6 hours in a week and a minimum of 12 weeks. In the "school experiences" course, candidates only make an observation to school environment, in-service teacher behaviors, his/her classroom management behaviors, and his/her teaching methods. Candidates do not have a chance to choose a particular subject to teach. In the "teaching practice" course, they have opportunity to teach school subjects (HEC, 2005, 2009). The decision of how many hours one candidate teaches depends on the universities, but a general perspective is one candidate should teach at least one hour in a week. However, in reality, candidates do not have the opportunity to teach every week, because of a heavy curriculum, crowded classrooms, national university entrance exam... etc. For these reasons, they have limited experiences for teaching science as well as the other subject matters.

This program was revoked by the HEC again. Students have not been able to enroll in upper secondary science teacher education programs since 2013. Most of the biology, physics, and Chemistry undergraduate programs were closed. This was a political decision made by the government. Despite the fact that there is a limited number of vacant positions in biology, physics, and chemistry subject areas, the Faculties of Arts and Sciences keep graduating people in many subject areas without making any needs analyses. This practice should be abolished immediately, because graduates are unable to find a job to contribute the society. On the contrary, with the Pedagogical Formation Certificate Program (PFCP), the government gives a chance to graduated students who later decide to become teachers for various reasons from non-education faculties. This program is not free. From time to time, this program was opened in order to close-up unfilled teaching positions. At the outset, this program was 4 semesters for graduate students. In time, it was diminished to 3 semesters then, in 2008, decreased 2 semesters. Meantime, it was also available for undergraduate students, having at least a 2.5 (depends on universities) GPA.

Since 1998 the PFCP has been known as the non-thesis master program. The students graduating from these programs are not given a masters degree. Instead they become qualified as teachers. This program was carried out in 3 semesters over 42 weeks and students had to take twelve core courses (total 39 credits) and 2 elective courses (total 6 credits). There were five compulsory courses—introduction to teaching profession, growth and learning, planning and evaluation in education, special teaching methods I, and school experience I—consisting of twelve hours of lecture and 8 hours of practice in a week that were taken in the first semester. The second semester also had five courses—instructional technologies and materials development, classroom management, special teaching methods II, and selective subject I—consisting of 10 hours of lecture and 10 hours of practice in a week. In the third semester there were four courses—analysis of study area course book, guidance, teaching practices, and selective subject II—consisting of 10 hours of lectures and 8 hours of practice in a week. This program gives people another opportunity to have a job and make money. But only 5–6% of the people have gotten this opportunity because of inadequate vacant teacher positions.

On the other hand, most of the PFCP teacher candidates don't love and don't value the teaching profession and have a negative attitude towards the teacher profession. These views may affect negatively their professional achievements and satisfaction. It is not a desirable situation both for themselves and their students (Basbay, Unver, & Bumen, 2009; Basturk, 2007; Eraslan & Cakıcı, 2011; Yuksel, 2004). This is very hard to overcome for academicians at the universities. Most of the academicians do not focus on or try to stop this problem. They have just lectured with conservative teaching methods and take extra money given for this program (Deniz & Sahin, 2006).

These days the PFCP is still in use in the Turkish education system, but it is 2 semesters long and only for graduates from universities. Although some courses' names were changed, PFCP students take parallel courses in education faculty programs. PFCP students have to take 10 courses. These courses are—introduction to education (2 credits), teaching principles and methods (2 credits), measurement and evaluation in education (2 credits), classroom management (2 credits), educational psychology (2 credits), instructional technologies and material design (3 credits), special teaching methods (3 credits), teaching practice (5 credits); and three elective courses which are guidance (2 credits), program development in education (2 credits), and sociology (2 credits).

Upon successful completion of the undergraduate program within eight semesters, Turkish teacher candidates receive lower or upper secondary school teaching licensure. This licensure can be used only at private schools, not public schools. The only way to start a teaching career by this licensure in public schools is to pass the national Public Personnel Selection Exam, called in Turkish KPSS.

Until 1985, the appointment of teacher candidates did not require a selection process; they could work in public schools because there was an absence in Turkish schools. After 1985, the government started to implement the Teacher Proficiency Exam, a preliminary-selection exam organized between 1985 and 1991 by the MoNE for the appointment of teachers and the Civil Service Exam, a central preliminary and selection examination held by the state in 1999 for appointment to any kind of state institution (including teacher appointments; Gündoğdu, Çimen, & Turan, 2008). In 2001, the Public Professional Exam replaced the Civil Service Exam. This is a central selection examination held by the Student Selection and Placement Center, called in Turkish OSYM for appointment to any kind of state institution including teacher appointments.

In 2002, KPSS was put into effect. The aims of KPSS are the selection of personnel for first-time public service duties and pre-selection of personnel to be assigned to public sector positions and bodies through special talent tests. KPSS scores are valid for 2 years. For the teacher candidates (see Table 6.4), the KPSS is held in two sessions on Saturday morning and Saturday afternoon. The session on Saturday morning is composed of the General Ability and General Cultural Knowledge tests, including 120 questions. It takes 120 minutes to complete the test. The Saturday afternoon session is composed of the Educational Sciences test, including 80 questions and takes 100 minutes. The following weekend, there is another exam, the Science test, including 50 questions and takes 75 minutes (OSYM, 2015).

The KPSS is highly significant for teacher candidates from a psychological, economic, and social perspective. KPSS system decides the region of appointment, the capacity of the class, and the circumstances under which s/he will work. In addition, what worries teacher candidates more is to be appointed as a permanent staff or on contract as soon as possible, and passing successfully the KPSS to this end (Özsarı, 2008). For a teacher candidate, not being able to begin his/her professional career at an educational institution results in, from an economic perspective, not earning money for at least one academic term; from a psychological perspective, failure at the end of a long and laborious educational life; and from a social perspective, lack of insurance or being considered from outsiders as an unemployed person despite his/her university (Üstüner, 2004). The teacher candidates who do not want to go through all these negative experiences are enrolled in courses designed to prepare them to pass the KPSS, postpone pursuing their interests, limit their social life, reduce the time spent with friends, and put significant effort into getting high scores from the KPSS. However, the work is not over with these efforts; the teacher candidates need to perform above a certain score in order to be appointed and make it to the quota in the ranking.

TABLE 6.4 KPSS for Lower/Upper Secondary Science Teachers

<table>
<tr><td rowspan="2">KPSS</td><td>Session I</td><td>General Ability
60 questions, 60 minutes
General Cultural Knowledge
60 questions, 60 minutes
(Turkish language 30, Math 27, Geometry 3, History 27, Geography 18, Citizenship 15)</td><td>Educational Sciences
80 questions, 100 minutes</td></tr>
<tr><td>Session II</td><td colspan="2">Pure Science (Biology, Chemistry, Physics)
50 questions, 75 min.</td></tr>
</table>

CONCLUSION

The Turkish education system has not had a stable order for a few decades. HEC has restructured all of the teacher education programs that currently exist in Turkey, and continue to develop and modify them as necessary along with the aim of entering the EU. However, there are a number of challenges, problems, and issues that make teaching science from regional, global, and multicultural perspectives necessary. The east part of Turkey has more problems than the west part considering economy, culture, and life conditions. These reasons cause unequal quality of public schools. Rich families prefer to send their children to private schools or take extra private lessons for their children. Most of the schools have physical condition problems, like a lack of laboratories, and educational technologies (computers, smart boards, projectors, etc.). Even some public schools do not have enough teachers to teach different disciplines. Overcrowded classrooms, shortage of teachers, and also an inadequate allocation of instructional resources are just a few issues. Besides inadequate educational issues, there are political problems with ongoing terrorism for 3 decades and refugees from Syria and Iraq. Due to terrorism, many families have to move from rural areas to urban districts or from less developed regions to more industrialized areas, or basically from the east to west part of Turkey. Lately, over two million Syrian and Iraqi refugees passed through the Turkish borderline and started to live in Turkey. These issues create an extra load on and difficulties to plan its educational facilities for the MoNE. Some village schools have been closed because there are no students left; in others many students have to be trained in integrated classrooms, with three or five grade levels sharing one room and a single teacher (Akarsu, 2000).

On the other hand Turkish secondary teachers are by far the lowest-paid of any OECD country. Because of these considerations, many teachers move on to other jobs when an opportunity arises. Therefore, these faculties have been among the lowest preferences of higher education entrants, so that they tend to have to accept the least-qualified entrants. This phenomenon stems from the lack of knowledge in students about the substance of teaching, reinforced by the low pay and social status of teachers.

In the information and technology age, teachers of the future should possess in professional competence at his/her study subject area, methodology, teaching, and teaching techniques (Bilgen, 1999). Most science teachers need a comprehensive professional development through both their pre-service and in-service periods on how to apply it in the classroom, how to design the lessons, and activities that support this kind of learning. However, the most important obstacle in implementation of the curriculum is teacher perception. Hansen and Olson (1996) state that most science teachers consider teaching the principles of science as their most important task and hesitate to change their teaching in any way. Understanding the belief and value structures of teachers is essential to improving their professional careers. Beliefs and perceptions are the most valuable constructs which shapes teachers' instructional design (Elmas et al., 2014; Ozden, 2007). These beliefs and perceptions are the driving force that shape who they are as teachers. Without overcoming the belief structures which shapes their perceptions, persuasion of teachers with the current curriculum ideals is not possible. Therefore, focusing on perceptions and opinions of the teachers is critically important, and it is a prerequisite for a long term influential change (Elmas et al., 2014).

A final important point is student's education life. The system is exam oriented, having negative effect on students' lives. Teachers have to focus much more on the transition exam from lower to upper secondary school (TEOG) or entrance of higher education exams (YGS and LYS), or public personnel selection exam (KPSS). Thus, it is hard to include student-centered strategies in their science lessons for teachers because of exam pressure. Students have to memorize so much science knowledge and have no time to be social, or engage in cultural activities. So, this life makes students not be open-minded; on the contrary, it makes them feel depressed and stressed, like a horse in race.

REFERENCE

Akarsu, F. (2000). *Transition and education: A case study of the process of change in Turkey.* In K. Mazurek, M. Winzer, & C. Majorek (Eds.), Education in a global society (pp. 315–327). Boston, MA: Allyn and Bacon.

Aksit, N. (2007). Educational reform in Turkey. *International Journal of Educational Development, 27*(2), 129–137.

Akyuz, Y. (2004). *History of Turkish education: From beginning to 2004* (9th ed). Ankara, Turkey: Ani Press.

Altan, M. Z. (1998). A call for change and pedagogy: A critical analysis of teacher education in Turkey. *European Journal of Education, 33*(4), 407–417

Basbay, M., Unver, G., & Bumen, N.T. (2009). A Longitudinal study on secondary education teacher candidates' attitudes towards teaching profession. *Educational Administration: Theory and Practice, 15*(59), 345–366.

Basturk, R. (2007). Investigation of test anxiety levels of pre-service teachers taking civil servant selection examination. *Fırat University Journal of Social Science, 17*(2), 163–176.

Bilgen, N. (1999). *Concerns and goals of Turkish high education at the outset of 21st century.* Ankara, Turkey: A Publication of Prime Ministry.

Bozyılmaz, B., & Kılıç, G. B. (2005). 4. ve 5. sınıf fen ve teknoloji dersi öğretim programının bilim okur-yazarlığı açısından analizi [Analyzing 4th and 5th science and technology instruction program by the science literacy aspect]. *Eğitimde Yansımalar: VIII Yeni İlköğretim Programlarını Değerlendirme Sempozyumu*

Bildiriler Kitabı (Reflecitons in Education: 8th Primary Education Programs Evaluation Symposium Proceedings) (pp. 320–328). Ankara, Turkey: Sim Press.

Çakır, Ö. (2005). Anadolu üniversitesi açıköğretim fakültesi ingilizce öğretmenliği lisans programı ve eğitim fakülteleri ingilizce öğretmenliği lisans programı öğrencilerinin mesleğe yönelik tutumları ve mesleki yeterlik algıları [Anadolu University open faculty and education faculties' English preservice teachers' attitudes towards teaching and professional self-efficacy perceptions]. *İnönü Üniversity Journal of Education Faculty, 6*(9), 27–42.

Çakıroglu, J., & Çakıroglu, E. (2003). Reflections on teacher education in Turkey. *European Journal of Teacher Education, 26*(2), 253–264.

Deniz, S., & Sahin, N. (2006). The Restructuring process of teacher training system in Turkey: A model of teacher training based on post-graduate education (PGCE). *Journal of Social Sciences, 2*(1), 21–26.

Elmas, R., Öztürk, N., Irmak, M., & Cobern, W. W. (2014). An investigation of teacher response to national science curriculum reforms in Turkey. *Eurasian Journal of physics & Chemistry Education, 6*(1), 2–33.

Eraslan, L., & Cakıcı, D. (2011). Pedagogical formation program students' attitudes towards teaching profession. *Kastamonu Education Journal, 19*(2), 427–438.

Erdogan, M. (2005). *İlköğretim 7. sınıf öğrencilerinin atomun yapısı konusundaki başarılarına, kavramsal değişimlerine, bilimsel süreç becerilerine ve fene karşı tutumlarına sorgulayıcı-araştırmacı yönteminin etkisi* [The effects of Inquiry method on middle school (level 7) students' achievement, conceptual change, scientific process skills and attitude toward science about the atomic structure]. Unpublished Master Thesis. Gazi University, Ankara.

Gokmenoglu, T., & Eret, E. (2011). Curriculum development in Turkey from the viewpoints of research assistants of curriculum and instruction department. *Elementary Education Online, 10*(2), 667–681.

Grossman, G. M., Onkol, P. E., & Sands, M. (2007). Curriculum reform in Turkish teacher education: Attitudes of teacher educators towards change in an EU candidate nation. *International Journal of Educational Development, 27(2),* 138–150.

Grossman, G. M., & Sands, M., & Brittingham, B. (2010). Teacher education accreditation in Turkey: The creation of a culture of quality. *International Journal of Educational Development, 30,* 102–109.

Gun, F., & Baskan, G. A. (2013). New education system in Turkey (4 + 4 + 4): A Critical Outlook. *Procedia—Social and Behavioral Sciences, 131,* 229–235.

Gündoğdu, K., Çimen, N., & Turan, S. (2008). Perceptions of prospective teachers in relation to civil servant selection exam (KPSS). *Journal of Ahi Evran University Kırşehir Education Faculty, 9*(2), 35–43.

Guven, I. (2008). Teacher education reform and international globalization hegemony: Issues and challenges in Turkish teacher education. *International Journal of Social Sciences, 3*(1), 1–17.

Hansen, K.-H., & Olson, J. (1996). *How teachers construe curriculum integration: the Science, Technology, Society (sts) movement as Bildung. Journal of Curriculum Studies 28*(6), 669–682.

Higher Education Council. (1999). *Teacher Preparation.* Retrieved from http://www.yok.gov.tr/web/guest/yayinlarimiz

Higher Education Council. (2005). Eğitim Fakultelerinde Uygulanacak Yeni Programlar Hakkinda Aciklama. TC Yuksek Ogretim Kurumu Baskanligi, Ankara, 2005. Retrieved from http://www.yok.gov.tr/documents/10279/49665/aciklama_programlar/aa7bd091-9328-4df7-aafa-2b99edb6872f

Higher Education Council. (2009). *Üniversiteler.* Retrieved from http://www.yok.gov.tr/content/view/527/222/lang,tr_TR

Irez, S. (2006). Are we prepared?: An assessment of preservice science teacher educators' beliefs about nature of science. *Science Education, 90*(6), 1113–1143.

Karadeniz, C. B. (2012). Opinions of teachers, Towards 4 + 4 + 4 Compulsory Education System. *Education Science Society Journal, 10*(40), 34–53.

Kavcar, C (2002) Cumhuriyet döneminde dal öğretmeni yetiştirme. *Anakara University, Journal of Education Sciences, 35*(1–2), 1–14.

Koc, Y., Isiksal, M., & Bulut, S. (2007). Elementary school curriculum reform in Turkey. *International Education Journal, 8*(1), 30–39.

Memişoglu, S. P., & Ismetoglu, M. (2013). The school administrators' conceptions concerning the system of 4+4+4 in compulsory education. *Journal of Research in Education and Teaching, 2*(2), 14–25.

Ministry of National Education. (2004). *İlköğretim fen ve teknoloji dersi (4. ve 5. sınıflar) öğretim programı* [Science and technology teaching program (4th and 5th grades)]. Talim ve Terbiye Kurulu Başkanlığı, Ankara.

Ministry of National Education. (2006). *İlköğretim fen ve teknoloji dersi (6, 7 ve 8. siniflar) öğretim programı* [Science and technology teaching program (6th, 7th and 8th grades)]. Talim ve Terbiye Kurulu Başkanlığı, Ankara.

Ministry of National Education. (2011a). *Ortaöğretim biyoloji dersi öğretim programı (9, 10, 11, ve 12. sınıflar)* [Secondary education biology teaching program (9th, 10th, 11th and 12th grades)]. Talim ve Terbiye Kurulu Baskanligi, Ankara.

Ministry of National Education. (2011b). *Ortaöğretim fizik dersi öğretim programı (9, 10, 11, ve 12. sınıflar)* [Secondary education physics teaching program (9th, 10th, 11th and 12th grades)]. Talim ve Terbiye Kurulu Başkanlığı, Ankara.

Ministry of National Education. (2011c). *Ortaöğretim kimya dersi öğretim programı (9, 10, 11, ve 12. sınıflar)* [Secondary education chemistry teaching program (9th, 10th, 11th and 12th grades)]. Talim ve Terbiye Kurulu Başkanlığı, Ankara.

Ministry of National Education. (2013). *Fen bilimleri dersi öğretim programı (3, 4, 5, 6, 7, ve 8. sınıflar)* [Science teaching program (3rd, 4th, 5th, 6th, 7th, and 8th grades)]. Talim ve Terbiye Kurulu Başkanlığı, Ankara.

Ministry of National Education. (2015). 2014–15 Formal Education Statistics in Turkey. Retrieved from http://sgb.meb.gov.tr/istatistik/meb_istatistikleri_orgun_egitim_2014_2015.pdf

Organisation for Economic Co-operation and Development. (2013). *Education policy outlook: Turkey.* Retrieved rom http://www.oecd.org/edu/EDUCATION%20POLICY%20OUTLOOK%20TURKEY_EN.pdf

Öğrenci Seçme ve Yerleştirme Merkezi. (2015). Retrieved from http://www.osym.gov.tr/belge/1-23227/2015-osys-kilavuz-ve-basvuru-bilgileri.html

Ozden, M. (2007). Problems with science and technology education in Turkey. *Eurasia Journal of Mathematics, Science & Technology Education, 3*(2), 157–161.

Özsarı, I. (2008). *Future related anxiety levels and vocational expectations of fourth grade preservice teachers related to KPSS examination* [Unpublished master thesis]. Istanbul, Turkey: University Social Science Institute.

Project 2061. (n.d.). Retrieved from http://www.project2061.org/publications/sfaa/

Tarman, B. (2010). Global perspectives and challenges on teacher education in Turkey. *International Journal of Arts and Sciences, 3*(17), 78–96.

Turkmen, H. (2005). Examining the technological history of Turkey impacts on teaching Science. *Science Education International, 7*(2), 115–123.

Turkmen, L. (2007). The history of development Turkish elementary teacher education and the place of science courses in the curriculum. *Eurasia Journal of Mathematics, Science & Technology Education, 3*(4), 327–341.

Unal, F., & Unal, M. (2010). Türkiye'de ortaöğretim müfredatlarının gelişimi. *Sosyal Bilimler Araştırmaları Dergisi, 1*, 110–125.

Üstüner, M. (2004). Today's past and present problems of Turkish education system teacher training. *Journal of Inonu University Education Faculty, 5*(7), 1–15.

Yuksel, S. (2004). The effects of the non-thesis master's program on students' attitudes towards the teaching profession. *Uludag Eğitim Fakültesi Dergisi, 17*(2), 355–379.

Yuksel, S. (2008). The tension in Turkish teacher education: Conflict and controversy during the transformation from an academic approach to a professional approach. *Asia Pacific Education Review, 9*(3), 367–379.

CHAPTER 7

SCIENCE TEACHER EDUCATION

A Multi-Country Comparison of the Israeli System—Critical Look

Nir Orion
The Weizmann Institute of Science

The Israeli secondary education consists of two stages: the junior high school level grades 7th–9th (age 12–15) and the high school level grades 10th–12th (age 15–18). The teachers' preparation and certification for the junior high level and for the first year of high school (10th grade) is conducted in about 20 colleges of education scattered all over the country as part of the BEd (Bachelors of Education) studies. Most of the colleges of education provide science studies and science teachers' certifications.

Science teachers' certification programs for the high school level are conducted at most of the universities' research institutes of Israel. The university graduation program (BSc) is a 3-year track, and the teaching certification is an external program for the BSc studies. Students usually start

Model Science Teacher Preparation Programs, pages 151–162

their teachers' certificate studies during their third year of the BSc program and are expected to complete this program in two academic years.

There are no official "standards" for teachers' education, however, since the Ministry of Education is responsible for both academic and practical standards of teachers preparation for all ages and subjects. In addition, the Ministry of Education influences deeply the content and the extent of the programs for teacher preparation in the colleges of education and in the universities.

The higher education in Israel is allegedly sovereign, self-governed by the Council for Higher Education. This council of about 20 members is nominated by the president of the state, but the board is proposed to the president by the minister of education. Thus, the content of the teacher preparation is suggested by the Ministry of Education and approved by the Higher Education Council that also provides the budget for each program. As a result, there is only minor freedom for each college and university to develop their own programs, and there is a close similarity among the college programs, as well as the university programs.

DESCRIPTION OF SCIENCE TEACHERS' CERTIFICATION PROGRAMS

Unfortunately, there are only a few publications that describe some parts of the universities science teachers certification programs in Israel (none of them in English). Thus, in order to base this chapter on evidences I had to collect data by myself specifically for this chapter through various sources of information such as yearbooks of universities, interviews with professors who teach the various programs, interviews with graduates of various programs, and interviews with science coordinators from the Ministry of Education.

The 4-year BEd program in the colleges of education consists of 48 hours of science courses (12 hours per year) and 34 hours of the education courses which are divided into 18 hours on theoretical aspects, 8 hours on didactical aspects, and 10 hours of practical experience in schools. These programs provide a certificate for teaching general science courses in the elementary school, the junior high school, and in the first year of high school (10th grade). The extent of the science studies does not allow the graduate of those programs to specialize in the teaching of a specific science (e.g., earth science, chemistry, and biology), therefore, they are qualified to teach only general science courses.

The educational component of the universities' teaching certificate programs is about half that of the colleges of education programs. The universities' programs include a theoretical component and a practical program. The theoretical component takes place in the university. This component includes 5–6 semesterial courses, with each course composed of about 24 hours. There is some variation among the different universities, but all the programs include the same following courses: philosophy and history of education, psychology of education, theories of learning and cognition, assessment of students' learning, and introduction of science education.

The content of the theoretical courses of the teachers' certification programs depends on the lecturer of each course; however, despite some differences in focus and extent, the following elements are common contents included across the various programs and lecturers.

Philosophy and History of Education

A typical course surveys the five schools of philosophy of education: essentialism, progressivism, perennialism, existentialism, and behaviorism and tries to locate them within a historical perspective. The interviews with lecturers and graduates of the various programs, indicted few courses that try to make clear a connection between exemplary Israeli schools and their educational philosophies.

Theories of Learning and Cognition

A typical course surveys the cognitive and socio-cultural approaches with some reference to behaviorism. Within the cognitive approach it discusses Piaget's theory of development and learning—the basis of constructivism; development from an information processing perspective (Neo-Piagetian approaches), issues of memory, and learning; the "time" dimension of learning (Ausubel); and metacognition. Within the socio-cultural approach, it includes Vygotsky's theory of learning and the neo-vygotskian addenda.

Educational Psychology

A typical course surveys different approaches in developmental psychology theories of children in different domains and their role in school functioning. Some courses focus on theories of adolescence and social processes in the classroom.

Introduction of Science Education

A typical course focuses on topics such as: goals and reforms in science education; science teaching strategies; scientific literacy; cognitive and affective aspects in students' learning (e.g., misconceptions); and assessment of students including traditional knowledge tests (open questions and multiple choice) and alternative methods such as projects and portfolios.

Assessment of Students' Learning

> A typical course introduces basic concepts and principles in the field of classroom assessment. It reviews traditional as well as alternative methods for assessing students' learning and achievements, with special reference to secondary school science. It also focuses on the issue of validity and reliability of assessment tools such as, "closed" questions, "open" questions, a research project, and a portfolio.

While the theoretical component is taught in large classes of mixed science disciplines students, the practical component is conducted in relation to a specific discipline and includes two courses: A 2-semester course of 48 hours focusing on the didactics of a specific discipline (e.g., didactics of biology/earth science/chemistry/physics) and a school-based practice of 180 hours. However, in reality, many students can accomplish only part of the practical experiences, since they can devote only one day a week for this purpose and it is very difficult to accomplish the required number of hours during this period.

The graduates of the university teacher certification programs subsequently become high school physics/earth science/chemistry/biology teachers, but since most of the secondary schools in Israel are 6-year comprehensive schools, there is a mobility of those teachers between the high school and the junior high school. Most of the graduates of the college of education programs usually teach science in the elementary school, and some elect to become junior high school science teachers. Since their scientific background is quite basic and general, it is very rare that a BEd graduate will teach a specific science discipline in the high school.

Although there is a significant difference between the BEd studies and the MSc studies in terms of the depth of breadth of the science studies (much higher in the universities) and the pedagogical studies (higher in the colleges), the structure of both programs is quite similar. Both programs are characterized by dichotomy. The first dichotomy is the clear distinction between the science studies and the pedagogical studies. In both programs, the pedagogical studies presents proper methods for teaching specific science subject such as chemistry, biology, or physics, but at the same time, they are exposed in their science studies (chemistry or biology or physics) to the "frontal" memory-based traditional science teaching, which is exactly the opposite of what they learned in their pedagogical studies. The same dichotomy exists within the educational studies between the practical and theoretical aspects of the program. For example, while the practical courses emphasize active learning, constructivism, inquiry-based learning, etc., such collaborative teaching is rarely implemented in the theoretical courses such as cognition, educational psychology, or philosophy of education. The effectiveness of practical experience in schools is also a controversial aspect of both teacher preparation programs. The practice component in schools depends on the local school's experienced science teacher who coaches the teacher candidate. However, teaching experience is a necessary factor for being a mentor, but definitely not sufficient. Therefore, there is a variance of mentoring abilities among those schools tutors.

One of my research programs expounds on that dichotomies exist as a case study. From the university students' course evaluation data received for 10 years, 3 courses taught by the same senior teacher got constantly higher score than others. These courses are philosophy of education, learning environments and the didactics of a specific scientific discipline. That teacher suggested in my interview the unique characteristics of his educational idea as follows.

> Teacher: I don't lecture—I teach and they are not courses—they are workshops. It means that I try to model a correct teaching. I teach them exactly as I teach my high school class. My field of research is science education and therefore my lab is the classroom and my subjects of study are the students. In my school lab, I also educate the new teachers. Instead of lecturing them that lecturing is not an effective teaching technique, they join me and take an active part in the teaching process. At the end of each class we discuss and analyze their concrete experience. Then, the theory follows the concrete.

The teacher used progressive teaching methodology for his course "philosophy of education" as follows, which is an example of the challenge for making difference. Before the beginning of the first lesson for about 50 students, the teacher arranged the classroom by 10 tables with five chairs. After students find and sit their chair, teacher introduces himself and give their assignment for this session. The group members have to do self-introduction each other, to explain others why they decided to join this program, to share others' attitudes/feelings towards the schools they had learnt, and if they have children who already go to school, to tell the group how do their children feel at school and then to discuss whether they would like their children to spent their childhood and youth and the same schools like they went to. Finally, they have to choose one members that will present the discussion and their agreements and disagreements. This emotional discussion continues during the presentations of each group. In the summary, the teacher only points to the different ideas that reflect the variety of personalities and philosophies of life that each of carries. Then he adds the final project of the workshop is to design "The school of my children." In the next session, the teacher set the class by different 10 groups and through a jig-saw activity each group has to fill a rubric of about 10 characteristics of each of the following streams: Essentialism, Perennialism, Progressivism, Existentialism and Behaviorism. Each group receives a short description of one of the above streams and by sharing information and knowledge with the other groups they fill the rubric of characteristics. Finally they have to present one stream through a role play. At the end of this activity, the groups email him their rubrics. He merges them and emails all participants a final rubric. Their next assignment is to explore any school they choose to find the characteristics and to identify its educational philosophy by using the rubric. The students are invited to present their finding to the assembly while completing this assignment. However, since about 95% of the Israeli schools present the versions of the Essentialism and Perennialism educational streams that are based on behavioristic methods, the workshop includes the 3- to 4-hour field trips to two schools, one of the progressive stream and an open school that presents the existentialism stream. The students are asked to go around, enter the class, observe lessons, and talk with local students or teachers freely for about 2 hours. Then last part of the visit is an open discussion with the principal, few students and teachers of the school. The field trip is discussed and summarized during the following workshop session as the most important part of the workshop because they changed their mind and are able to realize the benefits of an open and liberal education. The teacher never set the assignments as an obligation, and mark who arrive to the sessions and the field trip. The teacher asks students to email him their grade and a short reflection of the workshop, also notes that he takes himself the right to challenge their grade, but only when he thinks that they graded themselves lower than he would grade them. But, they of course have the right to accept his challenge or to reject it.

From Certification to License

The certification is only the first stage for teaching licensing. This license is given by the Ministry of Education following one year of apprenticing. The apprenticing stage takes place while the teacher candidate is positioning in a school as a science teacher and involves two components: An experienced, local school mentor who guides and supports the teacher candidate during daily teaching, along with participation in a year-long workshop (one meeting a week) in an authorized academic institute (college or university). The teaching license is given following the evaluation of the principle of the school and the Ministry of Education inspector and includes an observation of class session given by the teacher candidate.

The academic competences of the teacher candidates are the only terms for acceptance and graduation of the certification programs, and the licensure is based on a short period performance. Both stages of certification and licensure are not involved with any professional diagnostic of the psychological personality suitability of a teacher candidate to work with children.

THE "PERMANENT CRISIS" IN SCIENCE EDUCATION

According to Buchnik, Nathan, and Rave (2014), the number of high school chemistry and physics teachers decreased from 1988 to 2010 by 18%. In addition, the average age of the secondary science teachers increased during the last two decades (from 43 years old in 1996 to 50 years old in 2012), indicating that there is a significant decline in the entrance of young teachers to the science teaching profession. Therefore, if there will be no change in this area, then within a decade the Israeli high schools will face a significant shortage of science teachers especially in chemistry and physics.

The slow pace of entry of young science teachers in Israeli high schools is reflected in the number of participants or rather the number of students who complete their teachers certification programs at universities.

Israel has six higher education institutions that offer programs for science teachers certificate. Four universities offer programs for certification in two scientific disciplines—biology and chemistry or physics (Tel Aviv University, the Hebrew University, Bar Ilan University, and Ben Gurion University). The Technion (an academic institution for the engineering fields) has a teachers certification for chemistry, physics, and biology graduates (Hazzan, Herskovitz and Dori, 2015) and the Weizmann Institute (a research institution for only master's and doctoral students) offers programs for all four scientific disciplines: biology, chemistry, physics, and the earth sciences.

As mentioned above, there is no publication that presents quantitative data on the number of students graduating the various certificate programs each year, but the interviews that were conducted specifically for this chapter with professors of the various programs allow to estimate the range from 80–100 as the overall number of students who complete the various programs for science teachers certification in Israeli high schools.

The programs of the Weizmann Institute and the Technion are the largest programs among the six higher education institutions mentioned above. Since 2010, the number of students who graduate from the science teachers certificate program of the Weizmann Institute is about 20 students (10 biology, 4 chemistry, 4 physics, and 2 Earth Sciences). The teachers certificate program of the Technion tripled itself since 2012, when they started to operate the "Views" program and since then each year about 30 students (10 biology, 10 chemistry, and 10 physics) receive the teaching certification. The "Views" program invites Technion graduates back to the Technion to study toward an additional bachelor's degree in its Department of Education in Science and Technology. The degree they earn includes a high school teaching certificate for STEM subjects in one of 8 tracks: math, physics, biology, chemistry, computer science, environmental sciences, electrical engineering, and mechanical engineering. Technion graduates enrolled in the Views program receive full study scholarships from the Technion for two years. Since the number of credits required to complete this degree is similar to that required for an MBA, the students study one day or two half-days a week for two years, like in MBA programs, and can continue working as scientists and engineers in the industry in parallel to their studies. Although the Technion graduates who are enrolled in the Views program receive full study scholarships from the Technion for two years, they are not required to commit to teaching in the education system (Hazzan, 2014).

The technical characteristics of the "Views" programs (open to graduates, no tuition and no commitment) are identical to those of the Weizmann program and this is the reason why those two programs are much more popular than the programs of the others four universities, which are not opened for graduates and include tuition fees. Thus, the main issue is not how many complete, but how many turn to the teaching profession in practice. There is no published data concerning this issue, but following the interviews with members of several programs and with science coordinators of this Ministry of Education the common estimation was that 40%–50% of the graduates use their teaching certificate in practice and become science teachers. Moreover, among those who enter the teaching profession, about a third drop out during their first five years of teaching (Shperling, 2015). In order to fill the shortfall of science teachers, the Ministry of Education initiates a career retraining program for academics (mainly engineers) towards teaching science (mainly physics teachers). However, the effectiveness of those programs is also quite limited. The main reason for the dropout of teachers is their difficulty to deal with students; negative attitudes towards school in general and towards their teaching subjects in particular. The teachers find themselves between the rock and the hard place. Above them, there is a system of control that dictates to them what to teach, how to teach it, as well as how much and when, while pushing them towards an assessment-oriented teaching. Beneath them, there are the students who mostly resist this kind of teaching. Thus, teachers find themselves in the middle and empty handed, since the teaching certificate programs have not equipped them with the tools to generate interest and curiosity for learning internally among students. Those programs provided them with theory, but not enough practical models to follow in the realities of school life.

The above notion is supported by the following representative citations, which a version of each of them came out in most of the interviews that were conducted with graduates of the teachers certification programs:

> ...They gave us the theory with some examples and expected us to translate it to our daily teaching. The theory is important no doubt, but sorry, I don't know how to translate it to my classes.

> ...I don't need all these theories. I need someone who will assist me in my classroom, in the real world and not in the sterilized environment of the university.

> ...They bring us all these academy experts. Usually, these experts lecture us like they use to in the university. Maybe, it is a good strategy for university students, but it is boring and I can't learn from that what to do in my class.

Surprisingly (or maybe not) the above findings and citations are identical to two studies that were conducted about 20 years ago separately within the science teachers certification program of the Hebrew University in Israel and in Keele University of UK (Orion & Thompson, 1996; Orion & Thompson, 1999).

The shortfall of science teachers is usually viewed as a causal phenomenon of the crisis in science education. "The crisis in science education" is a term that has been used extensively for at least a half of a century (Bybee, 1993 . However, a consistent long-standing phenomenon cannot be considered a crisis; rather, it is the normal status quo. Thus, the "crisis" of science education is actually its normal status (at least in Israel) with some up and down fluctuations. The low status of the science teaching profession in Israel can be related to status of the science education, but it is more likely to relate it to the overall low status of the teaching profession in Israel. This claim is based on three indications: (a) the high social status of the science profession in Israel, (b) the low social status of the teaching profession in Israel, and (c) the shortfall of teachers characterizes many of the schools' disciplines.

Science graduates do not tend to turn to the teaching profession because it is not exactly attractive, and this lack of attractiveness is a reflection of the traditional perception of schooling. Despite the inspiring progressive theories of education, most schools in most countries preserve and present an essentialism philosophy (Imig and Imig, 2006). Essentialism refers to the "traditional" or "Back to the Basics" approach to education. It is so named because it strives to instill students with the "essentials" of academic knowledge and character development. This socio-economic approach sees the child as raw material and the school as in a position to mold the child into an obedient citizen for the rulers, as well as a productive power for the capital owners. Moreover, essentialism is grounded in a conservative philosophy that accepts the social, political, and economic structure of society. Essentialists argue that schools should transmit the traditional values and knowledge that students need to become model citizens. Essentialists maintain that classrooms should be oriented around the teacher, who should instill such traditional virtues as respect for authority, perseverance, fidelity to duty, and to focus heavily on achievement test scores as a means of evaluating progress.

The essentialism philosophy schools go very well within hierarchical societies that dominate the culture of obedience to authority and total dedication to the workplace. However, in less formal cultures, like in Israel, the essentialism approach provokes a considerable resistance by many students putting the teacher's role in a difficult situation. Unfortunately, the teachers of the teaching certificate programs have grown within the essentialism approach and are not equipped with practical methods and technics to generate an internal motivation among students and do not exposed them to role models who taught them according to the progressive approaches. Unfortunately, the teaching certification programs described above are part of the problem and do not contribute to the solution (Brameld, 1950).

THE DESIRED REFORM IN SCIENCE TEACHERS' EDUCATION

The university-based programs for teachers preparation should contribute to discontinuing the vicious circle discussed above. However, this can only be accomplished by changing the programs' perception of the essence of teaching. The change should include the following components:

1. To understand and internalize that the action target of teachers is human beings and not books or content. Then, to understand that the learning mechanism of human beings, similar to other animals, is instinctive and therefore occurs in response to stimulation. However, unlike other organisms, in humans the main stimulus for learning is emotional, and the cognitive ability follows the emotional need.
2. To understand that the first phase of the learning process is emotional and this aspect is maintained throughout the learning process together with the cognitive aspect. Therefore the teaching profession belongs to the family of the psychotherapy professions and a teacher should be provided with therapeutic skills such as listening, empathy, acceptance, and emotional support. This understanding should be followed by a radical changing of the extent and scope of the science teachers' certification programs.
3. The radical change of the science teachers' certification programs should involve the adapting of a new paradigm for teaching

profession and teachers' preparation. It is suggested to adopt the methods of educating and training physicians or doctors (Orion & Thompson, 1999). For example, in university hospitals, the professors of medicine work to educate and train new doctors in particular ways. We might advocate the setting up of special (university) schools where experienced "professors" of the science teaching department and distinguished school staff will daily teach pupils in the presence of, and with the cooperation of, their student tutors. In such a situation, these gifted and experienced teachers would work with the new teachers very closely and eventually invite them to take the lead in guiding the learners.

4. The implementation of the above model is only one component of breaking the dichotomy between theory and practice. An additional component is that all "theoretical" courses will implement active learning methods. This should be the preferred standard for the science courses as well.
5. The perception of teaching as a kind of psychotherapy profession should influence the personal criteria for accepting candidates for the teachers certification programs and the selection process of those who finally receive the certification.

The implementation of the above components will change dramatically the social status of the teaching profession, the personality suitability of the student teachers to the teaching profession, and quality of teaching. These reforms will dramatically change the profile of science teaching and its attractiveness even under the essentialism regime.

REFERENCES

Brameld, T. (1950). *Patterns of educational philosophy.* Austin, TX: Holt, Rinehart, and Winston.

Bybee, W. R. (1993). *Reforming science education. Social Perspectives & Personal Reflections.* New York, NY: Teachers College Press.

Buchnik, Z., Nathan, O., & Rave, A. (2014). *Science and technology education in Israel: Selected indicators towards building a risk management strategy for the expected shortage of teachers for science and technology in high schools.* Technion, Haifa: Neaman Institute Publications (In Hebrew).

Hazzan, O. (2014). *A proactive approach to high school STEM education in Israel, tomorrow's professor eNewsletter 1337.* Retrieved from http://cgi.stanford.edu/~dept-ctl/cgi-bin/tomprof/enewsletter.php?msgno=1337.

Hazzan, O., Hershkovitz, O., & Dori, Y. (2015). "Views" at the Technion chemistry teachers certification with an emphasis on "facing the future." *About Chemistry, 26,* 13–21. (In Hebrew).

Imig, D. G., & Imig, S. R. (2006). The teacher effectiveness movement: How 80 years of essentialist control have shaped the teacher education profession. *Journal of Teacher Education, 57,* 167–180.

Orion, N., & Thompson, B. D. (1996). Changes in perceptions and attitude of pre-service post-graduates secondary science teachers students. *International Journal of Science Education, 18,* 577–599.

Orion, N., & Thompson, B. D. (1999). Changes in perceptions and attitude of pre-service post-graduates secondary science teachers students—A comparative study of Israeli and British programs. *Research in Science & Technological Education, 17,* 165–192.

Shperling, D. (2015). *Dropping of teachers around the world: Review.* Mofet Institution. Israel. (in Hebrew).

CHAPTER 8

CANADA

An Overview of Secondary Science Teacher Education Programs

Christine Tippett
University of Ottawa

Todd Milford
University of Victoria

Canada is, in a word, diverse. For example, it spans a huge geographical area and is one of the most ethnically diverse nations on earth. Yet, despite its geographic and cultural diversity, Canada does have a national identity as each province and territory has representation at the federal level. Education in Canada is a provincial or territorial responsibility, rather than national, and as a result, there are a great many differences in how education is planned and implemented across the country.

In the first section of this chapter, we provide an overview of science education in Canada. We then explore issues associated with teacher education programs and multiple pathways towards certification. To provide a snapshot of Canadian secondary science teacher education programs, we

Model Science Teacher Preparation Programs, pages 163–184

tabulated data from multiple sources, allowing us to visualize the existence and relationship of common aspects. We conclude with an overview of the state of science teacher education within Canada.

Canada is located in the northern part of North America and is bounded by the Pacific, Arctic, and Atlantic Oceans, with the United States immediately to the south. Its total land mass is approximately 10 million square kilometres which are divided into ten provinces (British Columbia, Alberta, Saskatchewan, Manitoba, Ontario, Quebec, New Brunswick, Nova Scotia, Prince Edward Island, and Newfoundland and Labrador) and three territories (Yukon, Northwest Territories, and Nunavut). Canada is the second largest country in the world, geographically, but 37th by population, with a population of 35.5 million people as of 2013 (Statistics Canada, 2013, see also Figure 8.1). Canada has two official languages, French and English, and is one of the most ethnically diverse nations in the world; results of a national survey indicated that the population included more than 200 ethnic origins, 13 of which exceeded one million (Statistics Canada, 2015) Additionally, Canada has one of the highest per-capita rates of immigration; as of 2011, 20.6% of the population were born outside of Canada's foreign-born population (Statistics Canada, 2011). These geographic, cultural, and demographic characteristics have combined to influence the development and implementation of Canada's education, and therefore, science education, system.

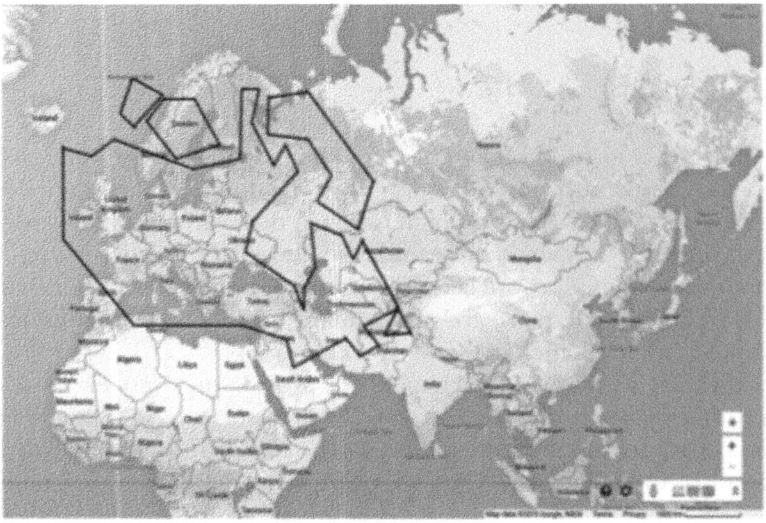

Figure 8.1 A rough outline of Canada superimposed upon a map of Europe (to scale) highlights the geographical challenges of the Canadian educational system (*Source:* adapted from Google Maps, 2015).

AN OVERVIEW OF THE CANADIAN EDUCATION SYSTEM

Most countries that traditionally score well on international assessments like the Trends in International Mathematics and Science Study (TIMSS) and the Program on International Student Achievement (PISA) have a national regulatory body that establishes guidelines for education and teacher education (e.g., Singapore, Japan, and Finland). In Canada, there is no such national body, and as a result, responsibility for education at elementary, secondary, and post-secondary levels falls to provincial and territorial departments, or ministries, of education. Thus, Canada has 13 educational systems with unique characteristics that have arisen from the varied geographic and cultural contexts.

While there are a great many similarities across the 13 provincial and territorial education systems, there are also significant differences. For example, as shown in Table 8.1, British Columbia has a School Act, which oversees public schools and the francophone education authority, a First Nations Education Act, and an Independent School Act which oversees schools outside of the first two acts. In Manitoba, the Public Schools Act authorizes the creation and regulation of the English and francophone school districts. In Ontario, the Education Act governs four publicly funded school systems: an English-language public school system, a French-language public school system, an English-language Catholic school system, and a French-language Catholic school system. Differences across these jurisdictions are highlighted in Table 8.1, which contains information about the department or ministry, legislation, and types of funded schools in each province and territory.

In Canada, free and accessible education is typically available from Kindergarten (age 5) to Grade 12 (age 17 to 18), with the mandatory attendance ranging from age 5 to 7 for entrance and 16 to 18 for exit. Common grade groupings are elementary, middle, and secondary. Specific grade configurations vary across and within jurisdictions and districts. For example, secondary schools could be Grades 7–12, 8–12, or 9–12. Very few Canadian secondary schools are designated as solely senior secondary schools with Grades 10–12 or 11–12.

Science Education in Canada

Although education is provincially and territorially mandated, resulting in 13 distinct systems of curriculum, the science curriculum is generally more cohesive because of the work of the Council of Ministers of Education, Canada (CMEC). CMEC is an intergovernmental body, which consults and provides information on a variety of issues relating to all levels of

TABLE 8.1 Governing Bodies, Legislation, and School Types Funded by Each Province and Territory[a]

Province or Territory	Governing Body	Legislation	Schools, as Identified by Province/Territory
British Columbia	Ministry of Education	School Act	Public (English) Francophone education authority
		First Nations Education Act	First Nations
		Independent School Act	Independent (incl. religious schools)
Alberta	Alberta Education	School Act (to be replaced by the Education Act in 2015)	Public (English, Francophone) Separate (incl. Catholic, Protestant) Charter
Saskatchewan	Saskatchewan Education	Education Act	Public Francophone Separate (Roman Catholic)
Manitoba	Education and Advanced Learning	Public Schools Act	Public (English, French Immersion, Français) Independent (incl. Catholic, Christian, Islamic, Jewish, Mennonite)
Ontario	Ministry of Education	Education Act	English Public English Catholic French French Catholic
Quebec	Ministère de l'Éducation, du Loisir et du Sport	Loi sur l'instruction publique	Public (French)
		Loi sur l'instruction publique pour les autochtones cris, inuit et naskapis	Cree, Inuit, and Naskapi
New Brunswick	Department of Education and Early Childhood Development	Education Act	Anglophone Sector Francophone Sector
Nova Scotia	Department of Education and Early Childhood Development	Education Act	English first and second language programs French second language programs Acadian French first language
Prince Edward Island	Department of Education and Early Childhood Development	School Act	English Language French Language

(continued)

TABLE 8.1 Governing Bodies, Legislation, and School Types Funded by Each Province and Territory[a] (continued)

Province or Territory	Governing Body	Legislation	Schools, as Identified by Province/Territory
Newfoundland and Labrador	Department of Education and Early Childhood Development	Schools Act	English Conseil Scolaire Francophone
Yukon Territories	Department of Education	Education Act	Public (English) French First Language Catholic
Northwest Territories	Department of Education, Culture and Employment	Education Act	English Public denominational school (Catholic) French Language education for francophone students
Nunavut	Department of Education	Education Act	Bilingual education (an Inuit language and either English or French) First language French

[a] This information was acquired from provincial or territorial government.

education across the country. One of its most influential initiatives was the *Common Framework for Science Outcomes K–12* (CMEC, 1997).

The *Common Framework* (CMEC, 1997) is significant as it was the first (and remains the only) effort to provide alignment of curriculum across Canada (Milford, Jagger, Yore, & Anderson, 2010). All provincial and territorial science curricula are based upon the *Common Framework* (CMEC, 1997) and a review of current Canadian science curriculum documents reflect that consistency. For example, *chemical reactions* is a topic mandated for all Grade 10 science students across the country. There is consistency, too, in the inclusion of science as a required subject from Kindergarten to Grade 10 across all provinces and territories. In Grades 11 and 12, students have the flexibility to specialize and can select from a wide variety of elective science courses (Milford & Tippett, in press).

CANADIAN TEACHER EDUCATION

Students pursuing a degree in education typically must select between elementary or secondary instructional levels (Crocker & Dibbon, 2008), although the range of options includes early childhood, primary, junior, intermediate, middle years, and adult education (Olson, Tippett, Milford, Ohana, & Clough, 2015). Additional options for specialization include content areas (e.g., English, mathematics, science) or focus areas (e.g., Aboriginal education, English Language Learners [ELLs], special education).

Pathways to Education Degrees

Teachers in Canada's publically funded schools typically require a minimum of a bachelor's degree in education (BEd) and a provincial teacher's certificate. There are multiple pathways for obtaining a BEd; at the 66 Canadian institutions that currently provide teacher education, more than 450 different programs exist (Association of Universities and Colleges of Canada, AUCC, 2014). However, across these multiple pathways, three main program structures have been identified: direct, consecutive, and concurrent (Olson et al., 2015). These structures exist for elementary, middle, and secondary teacher education, although options for direct entry to secondary science programs are limited. While none of the more than 450 programs can be considered typical, we randomly selected an example of a secondary science education program that follows each structure.

Direct Entry: University of Regina

The University of Regina's four-year program is one of few in Canada (outside Quebec) that permits direct entry from high school to a secondary education program. The program consists of 120 credits[1]and requires a 39 credit major (biology, chemistry, or physics) as well as an 18 credit minor, which could be a second science discipline. These credits include both science pedagogy and content courses. The program also includes at least three formal school placements, one of which is a 16-week practicum in a secondary school. Graduates of the program receive a BEd and are eligible to apply for a Saskatchewan Professional A Certificate, which entitles the holder to teach from pre-Kindergarten to Grade 12.

Consecutive Program: University of Victoria

The University of Victoria offers a post-degree professional program for undergraduate science degree holders who wish to teach science (biology, chemistry, physics, or general science) at the secondary level. Students complete 60 credits of pedagogy and foundation courses over a 16-month period to qualify for a BEd. Students have several field experiences, including a six-week practicum and an eight-week practicum. Graduates are eligible to apply for a BC Ministry of Education Professional Certificate, which is valid for any subject from Kindergarten through Grade 12.

Concurrent Program: University of Calgary

The University of Calgary's five-year program consists of three years with the Faculty of Science (Years 1, 2, and 4) and two years with the School of Education (Years 3 and 5). Applicants may enter directly from high school or with a maximum of 36 credits at the undergraduate level. Teachable areas are biology, chemistry, or physics and successful students graduate with a Bachelor of Science (BSc, natural sciences major) and a BEd. The program includes up to 9 credits of science teaching courses and 24 weeks of field experience. Upon completion of the program, students can apply for an Interim Professional Certificate, which is valid from Kindergarten to Grade 12.

Common Aspects of Canadian Teacher Education Systems

There are a number of factors that impact teacher education in Canada. The most significant factor is the lack of a national governing body for education, resulting in 13 provincial and territorial education systems. However, within each education system there are common aspects, for example, regulatory bodies that oversee accreditation of programs and certification of teachers. Another common aspect is that all systems have established the basic requirement of a university degree in education (Falkenberg, 2010). These common aspects and their relationships are shown in Figure 8.2 and explored in more detail in the sections that follow.

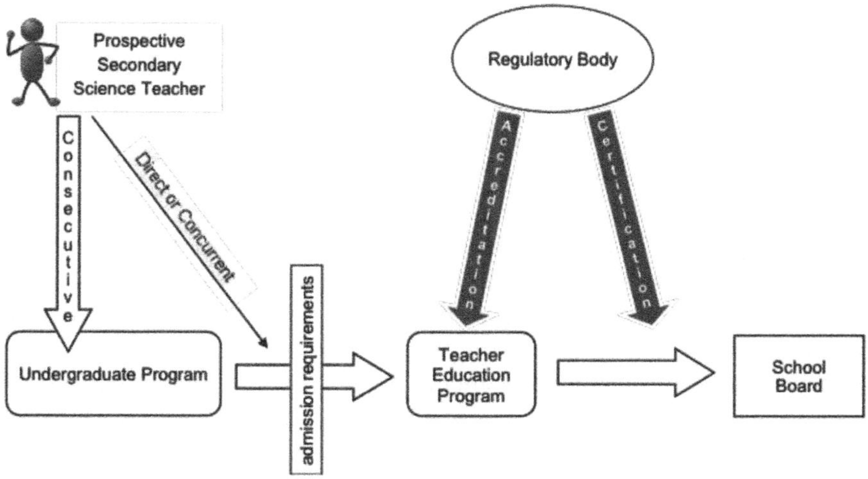

Figure 8.2 Common aspects of provincial teacher education systems.

Program Requirements

Because education is a provincial or territorial responsibility, there are no national program requirements either for teacher education in general or for secondary science teacher education in particular. It would be impossible to provide details about every program at each institution across the country, so to gain a sense of "national" approaches to secondary science teacher requirements, we selected a target university in each province. The territories' education programs are associated with specific provincial institutions: The Yukon and Nunavut work with the University of Regina while the Northwest Territories works with the University of Saskatchewan.

Our first choice was the university that bears the province's name; if there is no such institution, we selected the university with the largest BEd program. The consecutive structure is the most common across Canada, with the exception of Quebec (Crocker & Dibbon, 2008); therefore, in this section we highlight consecutive secondary teacher education. In the case of Quebec, teacher education programs are almost exclusively direct or concurrent (Crocker & Dibbon, 2008), making comparisons virtually meaningless. Accordingly, we omit Quebec in our cross-country analysis of secondary science education programs. We summarize admission requirements, pedagogy courses, pedagogical content knowledge (PCK) courses, and field experiences for our nine target universities. We do not address additional content requirements, because admission to these programs is contingent upon the successful completion of a specified number of university science courses.

Admission Requirements. Universities set their own admission requirements to accommodate the particular pathways they offer; as a result, there is a great deal of variation even for secondary science education program requirements. However, as shown in Table 8.2, common components include minimum grades in a previous degree, completion of science courses, and statements of relevant experience. Language proficiency (French or English) is also often an explicit requirement.

Pedagogical Requirements. Once students are accepted into a secondary teacher education program, they are required to complete a combination of prescribed coursework and field experiences. Coursework typically includes foundations courses such as the history, philosophy, ethics, and sociology of education courses that are completed by all education students regardless of instructional level or specialization. Other common courses examine pedagogical aspects of literacy, assessment, psychological development, special education or exceptionalities, and Aboriginal education. These pedagogical requirements may be specialized according to instructional level, for example, secondary students would study adolescent development and elementary students would study child development. Pedagogical requirements for the

TABLE 8.2 Admission Requirements for Target Secondary Science Teacher Education Programs

Institution	Admission Requirements[a]
University of British Columbia	Undergraduate degree (equivalent to a UBC bachelor's degree) that includes: • 6 credits of English literature and composition • 6 credits of chemistry, 6 credits of mathematics, 6 credits of physics, 3 credits of earth science • 24 credits of biological sciences, chemistry, earth science, or physics (18 credits at the 300 level and above) • >65% average in upper level science courses
University of Alberta	3 or 4 year degree from an accredited post-secondary institution • English language proficiency • 18 credits of biology, general sciences or physics • 9 to 15 credits in second teaching area • GPA >2.0 (C or 65%)
University of Saskatchewan	60 credits of pre-requisite courses that include: • 6 credits of English and 3 credits of Indigenous studies • 24 credits from biology, chemistry, or physics • 15 credits from a second teaching area Additional requirements include: • An online interview • 35–40 hours working with children
University of Manitoba	Bachelor's degree that includes: • 30 credits of biology, chemistry, general science, or physics • 18 credits of teachable minor • GPA >2.5 (C+) Additional requirements include: • proficiency in written English • 3 reference letters
University of Ottawa[b]	Minimum 3 year undergraduate degree from a recognized university or other postsecondary institution • applicants to the Grade 4 to 10 program must have completed 18 credits in science • applicants to the Grade 7 to 12 program must have 30 credits in biology, chemistry, or physics plus 18 credits in a second teaching area (which need not be science) • average of >70% Additional requirements include: • proficiency in English (as necessary) • statement describing experiences with children and youth in formal and informal educational settings
University of New Brunswick	An undergraduate degree that includes: • 3 credits of university level English • 30 credits in biology, chemistry, physics, or environmental science • 18 credits in a second area • GPA >2.7 (B–)

(continued)

TABLE 8.2 Admission Requirements for Target Secondary Science Teacher Education Programs (continued)

Institution	Admission Requirements[a]
Mount Saint Vincent University[c]	An undergraduate degree that includes: • 30 credits in biology, chemistry, geology/earth science, or physics • 18 credits in a second teachable • GPA >B (76%) Additional requirements include: • 3 reference letters • letter demonstrating experience in areas related to teaching • written proficiency in English
University of Prince Edward Island	Undergraduate degree from an approved registered university that includes: • 6 credits in English, 3 of which are recommended to be in composition • 6 credits in mathematics or statistics • 42 credits in science (chemistry, biology, foods and nutrition, forestry, geology/earth sciences, health sciences, kinesiology, oceanography, environmental science, agriculture, and physics) • 18 credits in a second area other than science • average of >70%
Memorial University of Newfoundland	Undergraduate degree (recognized by Memorial University) that includes: • 36 credits in biochemistry, biology, chemistry, earth science, or physics • 24 credits in another academic discipline, which could be a second science • average of >65%

[a] According to university websites: http://educ.ubc.ca/ https://uofa.ualberta.ca/education http://www.usask.ca/education/ http://umanitoba.ca/education/ http://education.uottawa.ca/en http://www.unb.ca/fredericton/education/ http://www.msvu.ca/en/home/programsdepartments/education/default.aspx http://www.upei.ca/education/ http://www.mun.ca/undergrad/programs/st_johns/education.php

[b] selected in the absence of a provincial university; the largest BEd program in Ontario

[c] selected in the absence of a provincial university; the largest BEd program in Nova Scotia

consecutive secondary science teacher education programs at the target provincial universities are shown in Table 8.3.

PCK Requirements. In addition to general pedagogical coursework, teacher education students are also required to complete more specialized courses that emphasize PCK in specific content areas. For secondary science education students, these courses are known generically as "science methods courses." Across our target universities, these PCK courses vary from general to level and discipline specific (e.g., intermediate science methods or methods in secondary physics), as shown in Table 8.4.

Field Experiences. As mentioned previously, field experiences are a required component of all teacher education programs in Canada. These experiences always include practicum placements and may include alternative educational

TABLE 8.3 Pedagogical Requirements of Target Secondary Science Teacher Education Programs

Institution	Pedagogical Requirements[a]
University of British Columbia	• at least 18 credits, including Aboriginal education, assessment, human development, literacy practices, supportive school environments, and teaching ELLs
University of Alberta	• at least 18 credits, including Aboriginal education, adolescent development, educational technology, inclusive education, language and literacy, and theory and practice
University of Saskatchewan	• 15 credits, including applied literacies, curriculum, exceptional learners, learning and development, and place-based learning
University of Manitoba	• 18 credits, including principles of teaching, Aboriginal education, special education, language and literacy, teacher and technology, and inclusive education
University of Ottawa[b]	• at least 15 credits, including curriculum planning, implementation, and assessment; learning theories in inclusive classrooms; and teaching at the intermediate/senior level
University of New Brunswick	• 18 credits, including assessment, inclusionary practices, and teaching and learning theories
Mount Saint Vincent University[c]	• up to 18 credits including technology in education, evaluation, inclusive education, and learning and development
University of Prince Edward Island	at least 24 credits, including assessment and evaluation, classroom management, curriculum and planning, inclusive education, Indigenous education, learners and learning, and teaching ELLs
Memorial University of Newfoundland	18 credits, including adolescent learners, assessment, exceptional learners, learning difficulties, reading in the content areas, and teaching and learning environments

[a] According to university websites: http://educ.ubc.ca/ https://uofa.ualberta.ca/education http://www.usask.ca/education/ http://umanitoba.ca/education/ http://education.uottawa.ca/en http://www.unb.ca/fredericton/education/ http://www.msvu.ca/en/home/programsdepartments/education/default.aspx http://www.upei.ca/education/ http://www.mun.ca/undergrad/programs/st_johns/education.php
[b] selected in the absence of a provincial university; the largest BEd program in Ontario
[c] selected in the absence of a provincial university; the largest BEd program in Nova Scotia

experiences such as community service learning. Placements range from one day per week to 16 consecutive weeks and span the continuum from observation to independent teaching as students progress through their program. In order to capture this continuum more precisely, the field experiences shown in Table 8.5 are described using consistent terminology as follows: *observation* (no teaching requirements), *orientation* (minimal teaching requirements), *independent teaching,* and *community service* (alternative placements or volunteer opportunities).

Accreditation. Accreditation is an external accountability process that ensures university graduates will meet established standards. In Canada, universities

TABLE 8.4 PCK Requirements of Target Secondary Science Teacher Education Programs

Institution	PCK Requirements[a]
University of British Columbia	• 6 credits: general science (secondary)—curriculum and pedagogy
University of Alberta	• 6 credits: curriculum and teaching for secondary school (majors I and II)
University of Saskatchewan	• 3 credits: selected from methods for teaching science in secondary school, methods for teaching secondary life science or methods in physical sciences
University of Manitoba	• 9 credits: curriculum and instruction in science plus 2 of teaching biology, chemistry, computer science, general science
University of Ottawa[b]	• 4.5 to 9 credits depending on junior/intermediate/senior designation: general science at the intermediate level, approaches to teaching science, and for senior levels, 2 of teaching science at the senior division (biology, chemistry, or physics)
University of New Brunswick	• 9 credits, including introduction to science education and advanced studies in science education
Mount Saint Vincent University[c]	• 6 credits: curriculum practices in secondary science
University of Prince Edward Island	9 credits, including intermediate/senior science and STEM
Memorial University of Newfoundland	6 credits: teaching and learning of science in the intermediate and secondary school (includes a laboratory component)

[a] According to university websites: http://educ.ubc.ca/ https://uofa.ualberta.ca/education http://www.usask.ca/education/ http://umanitoba.ca/education/ http://education.uottawa.ca/en http://www.unb.ca/fredericton/education/ http://www.msvu.ca/en/home/programsdepartments/education/default.aspx http://www.upei.ca/education/ http://www.mun.ca/undergrad/programs/st_johns/education.php

[b] selected in the absence of a provincial university; the largest BEd program in Ontario

[c] selected in the absence of a provincial university; the largest BEd program in Nova Scotia

and/or Faculties of Education must be accredited in order to offer programs that result in a degree that will be recognized for teacher certification. To be accredited, teacher education programs must include coursework and field experience. The accreditation process is almost always overseen by a provincial governmental agency such as the Ministry of Education or Department of Education. The sole exception is in Ontario, where the process is overseen by

TABLE 8.5 Field Experiences Within Target Secondary Science Teacher Education Programs

Institution	Field Experiences[a]
University of British Columbia	• one day per week for 4 months, orientation • one 2 week placement, orientation • one 3 week placement, community service • one 10 week placement, substantial independent teaching
University of Alberta	• three days, observation • one 5 week placement, some independent teaching • one 9 week placement, substantial independent teaching
University of Saskatchewan	• 2 days a week for 32 weeks, observation and some independent teaching • one 16 week placement, substantial independent teaching
University of Manitob[a]	• one 6 week placement, orientation • one 6 week placement, some independent teaching • one 6 week placement, independent teaching • one 6 week placement, substantial independent teaching
University of Ottawab	• one day per week for 15 weeks, community service • one 2 week placement and one 3 week placement, community service • one 8 week placement, moderate independent teaching • one 6 week placement, substantial independent teaching
University of New Brunswick	• one 2 week placement, orientation • one day per week for 3 months, some independent teaching • one 3 week placement, moderate independent teaching • one day per week for 3 months, moderate independent teaching • one 7 week placement, substantial independent teaching
Mount Saint Vincent University[c]	• four days, observation • one 2 week placement, some independent teaching • one 5 week placement, moderate independent teaching • one 2 week placement, some independent teaching • one 15 week placement, substantial independent teaching
University of Prince Edward Island	• six days, observation • one 8 week placement, orientation and moderate independent teaching • one 11 week placement, substantial independent teaching
Memorial University of Newfoundland	• one 2 week placement, observation and orientation • one 12 week placement, observation and substantial independent teaching

[a] According to university websites: http://educ.ubc.ca/ https://uofa.ualberta.ca/education http://www.usask.ca/education/ http://umanitoba.ca/education/ http://education.uottawa.ca/en http://www.unb.ca/fredericton/education/ http://www.msvu.ca/en/home/programsdepartments/education/default.aspx http://www.upei.ca/education/ http://www.mun.ca/undergrad/programs/st_johns/education.php

[b] selected in the absence of a provincial university; the largest BEd program in Ontario

[c] selected in the absence of a provincial university; the largest BEd program in Nova Scotia

the Ontario College of Teachers, which is a professional organization. Specific standards and program requirements vary according to province.

Certification. Certification is a means to regulate entry into the teaching profession, what competencies and standards must be met, and where teacher education may occur (Young & Boyd, 2010). As shown in Table 8.6, certification sometimes also determines what grade range an individual is permitted to teach. All provinces and territories require a BEd, and some offer monetary incentives for higher levels of education. Tests for teachers (either as entrance requirements or as part of certification) are rare to non-existent—we could not identify any jurisdiction that required such a thing at the time of this writing.

Teacher certification is usually overseen by the provincial Department or Ministry of Education. Ontario is an exception, where the Ontario College of Teachers regulates teacher certification. Despite the fact that teacher education programs are structured according to grade ranges (e.g., elementary versus secondary), certification typically includes K–12. Only three provinces (Ontario, Nova Scotia, and Newfoundland and Labrador) designate grade range restrictions for teaching certificates; although

TABLE 8.6 Granting Bodies and Certification Details, by Province and Territory

Province	Granting Body	Type	Grade Range
British Columbia	Ministry of Education	Certificate of Qualification	K–12
Alberta	Department of Education	Interim Professional Certificate	K–12
Saskatchewan	Ministry of Education	Professional A Teachers Certificate	Pre-K–12
Manitoba	Education and Advanced Learning	Professional Teaching Certificate	K–12
Ontario	Ontario College of Teachers	Certificate of Qualification and Registration	K-6, 4–10, or 7–12
New Brunswick	Department of Education	Initial Teacher's Certificate	K–12
Nova Scotia	Education and Early Development	Initial Teacher's Certificate	K-6 or 7–12
PEI	Education, Early Learning and Culture	Teacher's Certificate	K–12
Newfoundland and Labrador	Education and Early Childhood Development	Interim Teaching Certificate	K–6 or 7–12
Yukon	Department of Education	Professional Teaching Certificate	K–12
Northwest Territories	Department of Education, Culture and Employment	Teaching Certificate	K–12
Nunavut	Department of Education	Teaching Certificate	K–12

some provinces do note the grade range of the program completed, those notes are typically not restrictions. Some provinces recognize additional qualifications (e.g., Ontario); however, these qualifications are typically not mandated by school acts. Although secondary teacher education programs require teachable areas, those benchmarks are considered separately from certification (e.g., as teaching qualifications) and might only be considered if staffing is a highly competitive process (i.e., greater demand than positions available). Two provinces (Alberta, and Newfoundland and Labrador) grant interim certification, which can become permanent after two years of successful teaching. Table 8.6 provides information about the types of certification available for BEd graduates in each province and shows the grade range for which such certificates are valid.

Strengths and Areas for Improvements for Science Teacher Preparation

Our review of the secondary science education programs at the target universities across Canada revealed a variety of options available for potential secondary science teachers that offer a good degree of similarity and consistency but also hold unique and noteworthy differences. There is a common overarching structure (e.g., a BEd is a requirement for teachers in all provinces and territories, required coursework includes the categories of foundations, general pedagogy, PCK, and fieldwork) that affects admission requirements as well as the program components. Reflecting the categories of tabulated data in the sections above, here we discuss aspects of content knowledge, pedagogical knowledge, PCK, and fieldwork that are areas of strength and those aspects that might benefit from some modification, before we turn our attention to additional areas such as current curriculum revision, Canada's results in international assessments, and the lack of a national governing body for education.

In a 2014 report on 42 countries, the Organization for Economic Cooperation and Development (OECD) noted that the vast majority of secondary teacher program requirements include a placement (fieldwork or practicum) component as well as courses in content area subjects, and pedagogy. Additionally, two-thirds of these countries had programs that required courses in child/adolescent development and half of the countries required course work in research skills. According to our review, Canadian secondary science teacher programs closely fit the profile as described by the OECD (2014). However, there is variety in programs that is manifested in what specific courses are required in each of these common categories and how courses are implemented. Program requirements reflect the geographical and cultural contexts of the particular institution, and each course reflects those contexts, as well as the needs and interests of the particular instructor and students.

Content Knowledge

Each targeted university provides teacher education programs that address content, general pedagogy, PCK, and fieldwork. However, in the most common pathway to secondary science teaching in Canada, the consecutive program (Crocker & Dibbon, 2008), content knowledge is acquired at the undergraduate level and is addressed in admission requirements as shown in Table 8.2. The *Common Framework for Science Outcomes K–12* (CMEC, 1997) has clearly set out the content that is expected to be addressed in high school science. These expectations, in addition to the required undergraduate coursework in specific disciplines (e.g., biology, chemistry, earth science, and physics), ensure a level of consistency in content knowledge included in admission requirements for secondary science teacher education programs. Moreover, the admission requirements for Canadian education programs have been described as fairly stringent in comparison to many other countries (Crocker & Dibbon, 2008).

Pedagogical Knowledge

The courses in pedagogical knowledge that are offered within science teacher education programs across the country appear to be similar. For example, learning theory, assessment, special education, and classroom management are all core requirements; based on the data in Table 8.3, the number of credits required as well as course titles lead us to assert that the pedagogical knowledge requirements of Canada's science teacher education programs are comprehensive. In addition to these core courses in pedagogical knowledge, Canadian science teacher education programs also offer specific courses that reflect the country's history as well as its place within the global community. For example, Canada has a large indigenous population and a high rate of immigration. These factors have lead many programs to include courses in indigenous education as well as courses intended to address the needs of English language learners (ELLs). The educational needs of Canadian students are mirrored by the pedagogical requirements of the science teacher education programs at the target universities.

PCK

Park, Jang, Chen, and Jung (2011) have described PCK as "teachers' understandings of how to transform the content knowledge they have into forms that are pedagogically powerful and yet adaptive to variations in prior knowledge, understanding levels, and learning difficulties presented by individual students" (p. 246). The focus of most science methods courses can be assumed to be PCK, since these courses are often generically labeled "how to teach science." Data from the target universities suggests a wide range in the number of methods courses required, but even the most demanding programs include only 9 credits, or three courses that run for 36 hours each.

In the Canadian system, PCK is the one component of science teacher education programs that could be described as having room for improvement. According to Park et al. (2011), teacher PCK is significantly correlated with reform-based teaching (e.g., constructivist approaches, inquiry-based instruction) making it easy to justify devoting more time to enhancing the PCK of our pre-service science teachers. While some aspects of reform based teaching, such as classroom culture and teacher–student relationships (Sawada et al., 2002), could be fostered and developed in general pedagogy courses (e.g., classroom management courses), other aspects likely cannot. For example, content procedural knowledge or the design and implementation of inquiry units are highly discipline specific competencies (Park & Oliver, 2008; Sawada et al., 2000; Shulman, 1986).

Park and Oliver (2008) have conceptualized science PCK as five components: orientations to teaching science, knowledge of students' understanding in science, knowledge of science curriculum, knowledge of instructional strategies and representations for teaching science, and knowledge of assessment of science learning. Enhancing these components could be accomplished with additional science-teaching-focused field experiences or more time devoted to science methods courses. Even better might be integrated approaches in which secondary classroom experiences are embedded within science methods courses, bridging two contexts in which PCK can develop. Available information about the target universities' programs did not always reveal whether such an integrated approach was in place, but we know through firsthand experience and personal communication with colleagues that in many cases it is not. Emphasizing science PCK may also help new secondary science teachers respond to upcoming curriculum changes across the country.

Curriculum Changes—21st Century Skills

Despite the lack of a national science curriculum, the *Common Framework* (CMEC, 1997) continues to provide a common foundation for each province to build upon. This document was pivotal in broadening the scope of science education beyond simple content knowledge to include (a) science, technology, society, and the environment (STSE) aims and scientific skills, strategies; and (b) scientific habits of mind. However, the document was published nearly 20 years ago, and there have since been further reforms to science education on an international scale. 21st century skills (e.g., critical thinking, problem solving, creativity, innovation, social responsibility, leadership, collaboration, and communication, OECD, 2013;

The Partnership for 21st Century Skills, 2009) play a much larger role in current conceptions of what knowledge and competencies students need to succeed in post-graduation, and these skills are making an appearance in recent revisions to science curriculum documents in several provinces, including British Columbia and Alberta.

With a shift towards 21st century skills, PCK becomes even more vital as science teachers are expected to address higher level competencies and teaching methodologies that support or encourage problem based learning and inquiry. Science literacy is envisioned as incorporating 21st century skills along with content knowledge, emphasizing student-centered teaching of hands-on inquiry where the teacher is a facilitator rather than a dispenser of knowledge, while positioning language as the emphasis of the essential nature of language in doing and learning science (Aikenhead, 2007; Norris & Phillips, 2003; Tippett, 2011). It should be noted, however, that content must still hold a prominent position, since process and practices without content are not adequate to attain any measure of disciplinary literacy.

The content knowledge that prospective Canadian secondary science teachers hold can be considered comprehensive, judging by the target universities' admission requirements for secondary science teacher education programs. Most target programs include lengthy field experiences and a good range of general pedagogical courses. However, we argue that opportunities to develop PCK during those programs could be expanded. Overall, the composition of the secondary science teacher education programs at the target universities is robust; however, the impact of these programs is difficult to judge.

Teacher Assessment

In some countries, standardized assessment of teacher education program graduates is mandatory. For example, the United States has the Praxis Series, which for secondary science certification might include assessment of content knowledge in general science as well as specific disciplines; however, in Canada, standardized testing is not the overarching culture of assessment. In the absence of such measures for teachers, one way to assess how well secondary science teachers are doing in Canada is through regular (although infrequent) national and international evaluations of their students. Such assessments can serve as a proxy for content knowledge, pedagogical knowledge, and PCK. Canada, or provinces within Canada, currently participate in three such measures: (a) The Pan Canadian Assessment Program (PCAP), (b) the Trends in International Mathematics and Science Study (TIMSS), and (c) the Program for International School Assessment (PISA).

The CMEC is responsible, through the provincial and territorial ministries of education, for the PCAP. Taking place on a three-year cycle, PCAP is designed to assess the knowledge and skills of a representative sample of 13-year-old students from every province and territory in reading, mathematics, and science. On the most recent PCAP (2013), over 91% of Grade 8 students across Canada achieved the expected standards and 50% achieved higher levels of performance in science (CMEC, 2014).

A slightly broader perspective on the science literacy of Canadian students, and by extension, the quality of teaching, can be obtained through TIMSS, which measures the mathematics and science achievement of Grade 4 and Grade 8 students and takes place on a four- year cycle. Only some provinces choose to participate in TIMSS, and even those provinces that do, do not necessarily participate in each cycle. The average achievement score for all participating countries and regional jurisdictions is set at 500. On the 2011 iteration of TIMSS (Martin, Mullis, Foy, & Stanco, 2012), Grade 8 students from Alberta, Ontario, and Quebec scored 546, 521, and 520, respectively, well above the international average of 500. In 2007, Grade 8 students from Alberta, British Columbia, Ontario, and Quebec scored 543, 537, 536, and 517 respectively. Although two data point do not make a trend, Canadian Grade 8 students score well in science on TIMSS, a result that appears to be stable over the last two TIMSS cycles.

An even broader perspective on secondary science student achievement (and by extension some measure of secondary science teaching) is available from PISA. A representative sample of 15-year-old students in Canada is selected to participate in this assessment, which evaluates reading, mathematics, and science, and takes place on a three-year cycle. In 2012, Canadian students ranked 10th of the 65 countries that participated. PISA data can also reveal trends. Ranks of Canadian students over the previous four cycles were 8th (2009), 3rd (2006), 11th (2003), and 6th (2000). As is the case with the two other measures, Canadian students scored consistently well over all iterations of PISA. From these assessments, it is clear that Canadian secondary students are doing well (but certainly room for improvement remains) in science compared to national standards as well as international standards and comparisons to other countries.

As we have pointed out, there is no national regulatory body for education in Canada, which has implications for education at all levels, including teacher education programs. For example, there is no common prescribed curriculum across the country, although science education is a unique case because the *Common Framework* (CMEC, 1997) provides some degree of consistency. The decentralized nature of the Canadian education system also impacts teacher certification, along with both university and program accreditation.

Each jurisdiction has autonomy to establish and maintain their respective system of education, although the overall structure within each jurisdiction is similar. This lack of a centralized system is atypical among the nations that tend to score well on international measures of science achievement such as TIMSS and PISA. Perhaps the *Common Framework* for science curriculum provides a good balance between structure (recommended curriculum) and flexibility (nothing is prescribed), allowing teachers the pedagogical freedom to meet the learning needs of their particular students. This flexibility extends through the multiple levels of the education system in Canada (e.g., classroom, schools, districts, and university education programs).

CONCLUDING REMARKS

Canadian students do well on international assessment of science understanding, which reflects well upon Canadian secondary science teachers. The target secondary teacher education programs appear to be comprehensive in terms of content knowledge, general pedagogical knowledge, and field experiences. However, there does appear to be some room for improvements with regards to PCK, although it should be noted that all programs have PCK as a required element.

Canada is in a unique position within the global community in terms of secondary science teacher education. Unlike the other nations that do well on TIMSS and PISA, Canada is without a national curriculum. Instead, Canadian curriculum reflects regional and local needs and values within respective provincial and territorial educational systems. The *Common Framework* in Canada appears to provide the spine, fleshed out by the provinces and territories and made animate by the teachers. Perhaps this structured flexibility is what allows Canada to meet the diverse needs of secondary students and secondary pre-service teachers.

NOTE

1. Although various units are used across the country (including hours, credits, units, and credit hours) we have converted all program requirements to credit equivalents, where 3 credits is 36 contact hours, typically in a single term.

REFERENCES

Aikenhead, G. (May, 2007). *Expanding the research agenda for scientific literacy.* Retrieved from https://www.usask.ca/education/documents/profiles/aikenhead/expand-sl-res-agenda.pdf

Association of Universities and Colleges of Canada (AUCC). (2014). *University study programs in Canada.* Retrieved from http://www.aucc.ca/canadian-universities/study-programs/

Council of Ministers of Education, Canada (CMEC). (1997). *Common framework of science learning outcomes K to 12.* Retrieved from http://science.cmec.ca/

Council of Ministers of Education, Canada (CMEC). (2014). PCAP 2013: Report on the pan-Canadian assessment of science, reading, and mathematics. Retrieved from http://cmec.ca/Publications/Lists/Publications/Attachments/337/PCAP-2013-Public-Report-EN.pdf

Crocker, R., & Dibbon, D. (2008). *Teacher education in Canada: A baseline study.* Kelowna, BC: Society for the Advancement of Excellence in Education.

Falkenberg, T. (2010). Introduction: Central issues of field experiences in Canadian teacher education programs. In T. Falkenberg & H. Smits (Eds.), *Field experiences in the context of reform of Canadian teacher education programs* (pp. 1–50). Winnipeg, MB: Faculty of Education of the University of Manitoba.

Martin, M. O., Mullis, I. V. S., Foy, P., & Stanco, G. M. (2012). *TIMSS 2011 international results in science.* Chestnut Hill, MA: TIMSS & PIRLS International Study Center, Boston College. Retrieved from http://timss.bc.edu/timss2011/international-results-science.html

Milford, T. M., Jagger, S., Yore, L. D., & Anderson, J. O. (2010). National influences on science education reform in Canada. *Canadian Journal of Science, Mathematics, and Technology Education, 10*(4), 1–12. doi: 10.1080/14926156.2010.528827

Milford, T. M., & Tippett, C. D. (2016). Canada. In B. Vlaardingerbroek & N. Taylor (Eds.), *Teacher quality in upper secondary science education: International perspectives* (pp. 43–58). New York, NY: Palgrave Macmillan.

Norris, S. P., & Phillips, L. M. (2003). How literacy in its fundamental sense is central to scientific literacy. *Science Education, 87*(2), 224–240.

Organization for Economic Cooperation and Development. (2013). *OECD skills outlook 2013: First results from the survey of adult skills.* Retrieved from http://www.oecd-ilibrary.org/education/oecd-skills-outlook-2013_9789264204256-en

Organization for Economic Cooperation and Development. (2014). *Indicator D6: What does it take to become a teacher?* doi:10.1787/888933120252

Olson, J. K., Tippett, C. D., Milford, T. M., Ohana, C., & Clough, M. P. (2015). Science teacher preparation in a North American context. *Journal of Science Teacher Education, 26*(1), 7–28. doi:10.1007/s10972-014-9417-9

Park, S., Jang, J.-Y., Chen, Y.-C., & Jung, J. (2011). Is pedagogical content knowledge (PCK) necessary for reformed science teaching?: Evidence from an empirical study. *Research in Science Education, 41,* 245–260.

Park, S., & Oliver, J. S. (2008). Revisiting the conceptualization of pedagogical content knowledge (PCK): PCK as a conceptual tool to understand teachers as professionals. *Research in Science Education, 38*(3), 261–284.

Sawada, D., Piburn, M. D., Judson, E., Turley, J., Falconer, K., Benford, R., & Bloom, I. (2002). Measuring reform practices in science and mathematics classrooms: The Reformed Teaching Observation Protocol. *School Science and Mathematics, 102*(6), 245–253. doi:10.1111/j.1949-8594.2002.tb17883.x

Shulman, L. S. (1986). Those who understand: Knowledge growth in teaching. *Educational Researcher, 15*(2), 4–14.

Statistics Canada. (2013). *Population by year, by province and territory.* Retrieved from http://www5.statcan.gc.ca/cansim/a26?lang=eng&retrLang=eng&id=0510001&paSer=&pattern=&stByVal=1&p1=1&p2=37&tabMode=dataTable&csid=

Statistics Canada. (2015). *Ethnic origin reference guide, National Household Survey, 2011.* Retrieved from https://www12.statcan.gc.ca/nhs-enm/2011/ref/guides/99-010-x/99-010-x2011006-eng.cfm

The Partnership for 21st Century Skills. (2009). *P21 framework definitions.* Retrieved from http://www.p21.org/storage/documents/P21_Framework_Definitions.pdf

Tippett, C. D. (2011, September). *Scientific literacy: Disciplinary literacy and conceptual understanding.* Paper presented at the 8th biennial conference of the European Science Education Research Association, Lyon, France.

Young, J., & Boyd, K. (2010). More than servants of the state: The governance of initial teacher preparation in Canada in an era of school reform. *Alberta Journal of Educational Research, 56*(1), 1–18.

CHAPTER 9

NATIONAL REQUIREMENTS FOR SECONDARY SCIENCE PREPARATION

Elizabeth Allan
University of Central Oklahoma

The United States is a collection of independent states and territories for which there is no federal education system. States and municipalities have the authority to set standards, curricula, and criteria for enrollment and graduation. This applies to all levels of education, including higher education. The U.S. Department of Education (USDE) was established in 1867 with the primary mandate of collecting information about school systems and disseminating that information to provide guidance. The USDE continues that mandate today while extending its influence by using regulations tied to financial incentives for school systems. Financially, the majority of educational funding comes from municipalities and states. In 2014, the USDE provided approximately 11% of educational funding with the remainder coming from local municipalities and state-level funding (USDE, 2015b). However small the percentage appears, the impact is substantial to schools without strong local or state funding. It is through the regulation of these vital federal dollars that educational mandates direct the U.S. educational system. This includes federal impacts on educator preparation.

Model Science Teacher Preparation Programs, pages 185–203

THE HISTORY OF EDUCATOR PREPARATION REQUIREMENTS IN THE UNITED STATES

The long history of the U.S. education system and educator preparation is one of increasing standards, accountability, and control along decreasing perceptions of the value and worth of educators. Public education in the United States began in the 1800s with the origin of the common schools. Educator preparation emerged as a function of requirements for teaching in common schools. Most states put into place regulations that included academic competence and some limited training. As common education grew, the need for a trained educator workforce saw the emergence of the normal school. Beginning in Massachusetts in 1838, normal schools were to provide "normalized" systematic training to ensure a consistent quality of educators for public education (Lackney, 2015). While normal schools were the origin for educator preparation, the turn of the century saw educator preparation move to colleges and universities and the development of a systematic series of courses for teacher training. As the complexity of teacher training grew, so did the regulations and expectations.

Education, and educator training, has consistently been under fire in the United States. From Horace Mann and the common school movement to John Dewey and the progressive movement, public education has been the focus of calls for reforms, tightened standards, and greater accountability with educator preparation at the heart of each reform movement. More recently, it was the 1983 publication of a report A Nation at Risk (Garner, 1983) that portrayed teachers as unqualified, underpaid, and working in conditions that were not conducive to teacher or student success. The follow-up report in 1986, A Nation Prepared, became the foundation for professional requirements and a set of new standards for education. In 2001, The Elementary and Secondary Education Act was revised and renamed No Child Left Behind (2002). This legislation ushered into schools mandatory annual improvement for all subgroups of students tied to significant penalties if not achieved. All areas of education were addressed, including regulations for improving educator preparation and new requirements for student achievement. The 21st century standards movement had arrived (Minow, 2015).

Science Teacher Preparation Accountability

Accountability, defined here as a catchall phrase used to describe the measures taken by one agency to impact the decisions and actions of another agency, has been a consistent theme throughout the history of education and educator preparation in the United States. Science teacher preparation has not been exempt from accountability requirements. These requirements have helped shape science teacher preparation and continue to have an effect today. While this chapter is a review of the current state of science teacher preparation in the United States, no review would be complete without addressing the context within which science teacher preparation is found nationally.

The need for science teachers is at an all time high, and the shortage is felt nationwide. In the 2015 U.S. Department of Education (USDE) report of the national listing of teacher shortage areas, science was listed by every state as a high-need area (USDE, 2015c). Federal and state financial aid allocations reflect this need for more science teachers. Science is a high-need area for the TEACH grant program, the Federal Loan Forgiveness program, and is often supported by state grants or incentive programs (see http://www.okhighered.org/otc/tseip.shtml for one example). For example, the TEACH grant has dispersed almost 200 million dollars to 38,318 recipients (USDE, 2015a). These recipients were found in all 50 states from both private and public institutions. The National Science Foundation's Robert Noyce Teacher Scholarship program anticipates awarding $52,800,000 dollars to between 50 and 60 awardees in 2015 to increase the number of and/or the performance of science teachers (NSF, 2015). A great deal of public resources is being allocated to increase the number of science teachers. The quality of science teachers is also under national scrutiny.

The 2007 report, Rising Above the Gathering Storm (National Academy Press, 2007), showed trends in the lack of competitiveness of both students and teachers in science, technology, engineering, and mathematics education (STEM) in comparison to other industrialized nations. In response to growing concern over U.S. student performance, the Obama administration's competition, Race to the Top (RTTT), allocated four billion dollars to states that adopted policies and strategies to improve teacher quality. Since the origin of RTTT, there has been a nationwide move for greater teacher effectiveness marked by revamped and enhanced teacher preparation (CAEP, 2013) and teacher evaluation systems often using "value added measurements" (Marzano, 2012). The effectiveness of these movements has yet to be determined (Darling-Hammond & Rothman, 2015), but their impact is significant on science teacher preparation.

One such impact is the number of standards that science teacher preparation programs address. Science teacher preparation programs must meet state and national standards for teacher preparation (NSTA, 2012; CAEP, 2013). However, the standards affecting science teacher preparation are not just limited to those assessing the program itself. Programs must also prepare future science teachers to address state and national P–12 student standards such as the Next Generation Science Standards (NGSS Lead States, 2013) and National Governors Association Center for Best Practices and Council of Chief State School Officers (2010). States use these standards as the knowledge base upon which to build high-stakes P–12 student tests. High-stakes P–12 student testing is the norm for the United States, and a competent teacher plays a major role in creating an environment in which students can learn effectively. Evidence indicates that, overall, teacher preparation matters (Darling-Hammond, Holtzman, Gatlin, & Heilig, 2005). But it's hard to determine what makes for an effective program. While value added measurements of the performance of program graduates on the learning of P–12 students is one measure, it's become highly contested as a measurement of teacher effectiveness and thereby questionable as a measurement of the effectiveness of their science teacher preparation. Methodological difficulties arise when attempting to link student success to teacher preparation (National Research Council, 2010). However, some attempts have been made to determine what makes a quality science preparation program.

Presley and Coble (2012) determined four themes that represent the key characteristics of an ideal science and mathematics teacher preparation program. The National Research Council (NRC, 2010) provides evidence that supports some of the characteristics that are valuable for what science teachers should have, but the research is not conclusive on how teacher preparation programs can best develop those characteristics. Regardless of the research, one common theme is that a strong content knowledge and strong pedagogical content knowledge are recognized as critical for teacher success (NRC, 2010). How do science teachers acquire strong content and pedagogical knowledge? In the United States, the answer is not necessarily science teacher preparation programs.

Science teacher preparation programs, like other traditional preparation programs, are held accountable by state and national requirements. Every program must demonstrate program efficacy in order to be able to recommend graduates for state licensure. A major part of program reporting is providing evidence that graduates know their content and are able to teach it. The stakes for each program are high. Failure to adequately address accountability measures can result in the loss of the ability to recommend for state licensure and, in some cases, to be shut down. A large amount of energy and resources go to addressing the standards for educator preparation, and the demands on pre-service teachers are high. Given the evidence a traditional teacher preparation program provides about the quality of their graduates, why do other routes exist?

As the state and national requirements for science teacher preparation programs increase, high-stakes testing and teacher accountability measures act to drive increasing numbers of teachers out of the field. This results in a science teacher shortage at a time when it is increasingly difficult to recruit and retain high-quality candidates. Shortages result in an increasing number of teachers coming from alternative certification routes and an increased number of educators prepared outside of the traditional university or college (i.e., school systems, Teach for America).

Alternative certification is used by states with teacher shortages to put a teacher in every classroom when a traditionally prepared teacher is not available. Although alternative licensure represents a large number of licensed teachers, there is no definite definition of what it includes or excludes. The National Association for Alternative Certification (NAAC) defines alternative routes to certification or licensure as any preparation program other than a traditional undergraduate degree-granting program leading to certification or licensure. The use of alternative certification has increased dramatically with over 200,000 teachers estimated by the NAAC to have been certified through state-run programs (NAAC, 2015). Teach for America (TFA) and Troops to Teachers (TTT) are other well-known alternative routes to teaching that recruit from specific populations. TFA requires a bachelor's degree; a minimum 2.50 overall grade point average (GPA); legal citizenship, legal residency, or deferred action for childhood arrival status; and certification as required by the state they teach in. TFA members receive pedagogical training in a summer training program, coaching and mentoring from TFA staff, and have online resources available for them.

Troops to Teachers is a U.S. Department of Defense program that provides eligible service members financial support to become teachers. TTT participants must hold a baccalaureate degree when they apply and are required to complete the state's certification requirements (Troops to Teachers, 2015). These alternative routes start with individuals with an undergraduate degree and no pedagogical knowledge. And the number of alternatively certified science teachers is growing, with alternative programs preparing a higher proportion of STEM teachers than traditional programs (USDE, 2015c).

Teach for America had over 11,000 members for the 2013–2014 school year—up from 300 members in the 1990–1991 school year. Approximately 3,500 members teach in a STEM area (Teach for America, 2015). With around 227 thousand science teachers in the United States (NCES, 2014), TFA members reach a significant number of students.

Studies of the impacts of alternative routes on student learning are mixed. Heilig and Jez (2014) and Humphrey and Wechsler (2007) found wide variation between alternative certification programs making comparisons difficult. Alternative routes to science teaching provide classrooms with teachers who have a degree or a related degree in science. Pedagogical training is limited or non-existent in these programs. However, given the current state of education in the United States, they are not likely to be eliminated as routes to teaching science anytime soon.

Individual states set accountability measures and levels of competency, and as a result, there are no clear indicators or definitive levels of quality across the states. Information about programs within each state is available through the Title II report (USDE, 2014). As part of the reauthorization of the Higher Education Act of 1965, Title II (2014; Sections 205 through 208) requires states and institutions with teacher preparation programs to report Title II data to provide public access to data about teacher preparation and certification with each state. Data is provided for both traditional and alternative certification and is used to report to the U.S. Congress on the status of educator preparation. The report is a comprehensive picture of educator preparation in the Untied States and provides extensive data on both traditional and non-traditional programs. Programs designated by the state as low-performing are listed, and measures of quality are provided, (i.e., grants received, hours of clinical practices, test scores, etc.) but the report does not make quality rankings across states. The report is an important measure of the status of educator preparation in the United States but does not generally receive widespread reporting by the media.

While testing vendors are able to give statistics for each state and program, there are few, if any, measures or rankings across states. One such measure is the National Council on Teacher Quality (NCQT). The NCQT report, *2014 Teacher Prep Review* (2014) is the second of the series and provides a ranking of teacher preparation programs across the country. In the most current report, non-traditional teacher preparation programs (e.g., Teach for America) are included in the rankings. The report garnered great criticism for the methodology used to gather information, and as a result, less than 1% of preparation programs fully cooperated. The report bases program rankings on input measures such as published course requirements and course syllabi without measuring outputs such as the quality of instruction or the impact of graduates on K–12 student learning. While research does not support the report's findings (Darling-Hammond, & Rothman, 2015), and regardless of the criticism, the rankings and results are widely distributed for public consumption and have contributed to the perception of teachers as unprepared and teacher preparation programs as ineffective.

This chapter looks generally at the state of secondary science education and not at specific states, and does not make comparisons between states or programs. Additionally, traditional undergraduate science teacher preparation programs are evaluated at the program level and provide data on their graduates. It is this accountability and status of traditional science teacher preparation that is the focus of this chapter as similar information about alternative certification programs is not readily available.

NATIONAL ACCREDITATION FOR EDUCATOR PREPARATION

National accreditation for educator preparation began much earlier than the current call for reforming the teaching profession and the development of more rigorous standards for teacher preparation and student achievement. The National Council for Accreditation of Teacher Education (NCATE) was founded in 1954 as an independent nongovernment, national accrediting body of teacher education. NCATE and its partner organizations developed a set of standards used to evaluate schools, colleges, and departments of education termed "units" (NCATE, 2008).

NCATE furthered its impact on educator preparation by partnering with Specialized Professional Associations (SPAs) for the development of program-level competencies and standards for the review of individual programs of study. As part of the accreditation process, SPAs evaluated individual programs for sufficiently addressing individual program area competencies/standards. While every state and subject-specific program did not use the SPA standards, they were used a template for states to conduct their own subject area program reviews.

As a nongovernmental accreditation body, NCATE had no authority to require accreditation of educator preparation units and so developed partnerships agreements with all 50 states, the District of Columbia, and Puerto Rico. While state partnership agreements differed, twenty-three states required NCATE accreditation for all their educator preparation units, and thirty-one states required it for the majority (NCATE, 2015). Clearly, NCATE and the NCATE standards had a significant impact on the preparation of educators.

While not required by the federal government, NCATE was recognized as the sole accreditor of teacher education until 1997 when a second national accrediting association, the Teacher Education Accreditation Council (TEAC), was founded. Recognized in 2001 by the Council for Higher Education Accreditation (CHEA) and in 2003 by the USDE, TEAC differed significantly from NCATE philosophically. While both NCATE and TEAC provided national accreditation for the preparatory unit, NCATE required some form of external review of each program that prepared educators and developed a set of standards to assess the unit as a whole. Conversely, TEAC did not require program review and instead required an Inquiry Brief process in which the unit addressed three quality principles by having faculty in the unit present evidence of effectiveness. Accreditation was based on an audit of the evidence.

In 2009, NCATE and TEAC began the process to merge into a single national accrediting body and on July 1, 2013 CAEP became fully operational as the sole accrediting body for educator preparation providers. CAEP's standards are aligned with the Interstate New Teacher Assessment and Support Consortium (Council of Chief State School Officers [CCSSO], 2011) and emphasize content and pedagogical knowledge, partnerships with K12 stakeholders, selectivity of candidates, and the impact program's graduates have on K12 student learning. Evidence for programs meeting the standards rely heavily on assessments as part of a required quality assurance system (CAEP, 2015).

CAEP has three options for subject area program review: Option 1: SPA program review with National Recognition; Option 2: CAEP program review with Feedback; and, Option 3: State Program review (CAEP, 2013). Only Option 1 results in National Recognition, but all three options require evidence of the program candidate's knowledge, skills, and disposition. CAEP provides the pathways to program review, but states retain the authority to designate which pathway and who evaluates programs. As partnership agreements are developed with CAEP, the options available to evaluate individual secondary science programs will vary; however, it is the state that approves each secondary science program and grants teacher preparation programs the authority to recommend graduates for licensure in the state.

Secondary Science Teacher Preparation Standards

The National Science Teachers Association (NSTA), in collaboration with NCATE, developed the first set of NSTA Standards for Science Teacher Preparation in 2005. The standards were built using a research-based knowledge base (NSTE, 2012) and represent current best practices in the field of science education. While the most recent revisions (2012) were aligned with *A Framework for K–12 Science Education: Practices, Crosscutting Concepts, and Core Ideas* (Council, 2012) in anticipation of the NGSS, the standards are comprehensive enough to include expectations for all the required knowledge for beginning science teachers (Veal and Allan, 2014). Content knowledge, pedagogical knowledge, the learning environment, safety, impact on student learning, and professional knowledge and skills are the emphasized by the 2012 standards. As such, the standards are the only national set of expectations for what a beginning science teacher should know and be able to do. Used as described above in the program approval process by some states, the standards set strong expectations for strong content and pedagogy.

REQUIREMENTS FOR SCIENCE TEACHER PREPARATION

Part of the process of becoming a science teacher is passing entrance and exit exams for each program. As with other educational policies, states determine whether or not exams are required for programs licensed within the state. States determine both the exam used and the passing score for each exam. Secondary science education programs within each state must meet their state requirements for testing to be able to recommend their graduates for state licensure. These exams are also used in accreditation as evidence of candidate content knowledge.

While there are no nationally required exams, there are some commonalities among all states. In general, to gain admission to a teacher education program, candidates are tested on basic skills in reading, writing, and mathematics. To graduate from a preparation program and to be licensed, candidates are tested to measure specific subject content knowledge and subject-specific teaching skills. While some states create their own tests, the most commonly used are the Praxis Series Tests (ETS, 2015).

A final measure of candidate effectiveness is usually some form of assessment of their readiness to enter the teaching profession. This evaluation typically occurs during the final clinical experience and is a performance-based, subject-specific portfolio of artifacts developed by the candidate in response to a set of prompts. Commonly termed the Teacher Work Sample, several models are used across states (Girod, 2002). Commercially available performance assessments include the *Praxis* Performance Assessment for Teachers (ETS, 2015) and the edTPA (AACTE, 2012). These entry-level assessments focus on basic skills and knowledge and are meant to be comparable to other professional licensing exams such as medical licensing or the bar exam for lawyers. Each measures a candidate's content pedagogy, or more specifically the ability to plan, implement, and assess lessons to ensure all students learn. As noted above, each state determines what tests to use and the passing scores on each test. This makes comparisons between states difficult; however, the Title II report (USDE, 2014) outlines each state's requirements and provides rankings within each state for each institution and is part of the measure for low-performing institutions within each state.

While some alternative licensure programs require candidates to complete some measure of coursework, most states require alternative licensure candidates to pass the content knowledge exam, and in some cases, the required exam for subject-specific teaching skills (USDE, 2015c). This results in a significant imbalance of requirements between those teachers who go through a teacher preparation program and those that are licensed via an alternative route. One significant effect of this imbalance may be one reason for the reduction in the number of candidates in traditional preparation programs from previous years (USDE, 2015c).

SCIENCE TEACHER LICENSURE

There are several levels of certification or licensure within the United States. Licensure can be at an initial or advanced level, be for one subject or many, and license teachers to teach at multiple levels or restricted grades. To understand licensure in the United States is necessary to understand that licensure is required for every subject and grade in public schools.

Federal law, as part of the NCLB regulations for highly qualified teachers, requires that educators hold a bachelor's degree, demonstrate competency in the subject they are teaching, and hold full state certification or licensure. Just as is the case with most other decisions regarding education, each state has the authority to determine the levels of licensure for science educators. In general, states license at the initial and advanced level. The initial licensure is the first teaching license obtained but is not necessarily only at the baccalaureate level. Some programs are post-baccalaureate but lead to an initial license.

The route to initial license is determined by the competencies and testing requirements set forth by each state. Programs that are approved to recommend their graduates for licensure have demonstrated to the state that their program requirements (i.e., courses, clinical experiences, etc.) produce graduates that have met the state's requirements. Advanced licensure (e.g., principal, school counselor, etc.) is generally post-baccalaureate and requires an advanced degree or sequence of graduate courses. Licensure may also be dependent on passing an exam set by the professional organization or state. For example, the Council for Accreditation of Counseling and Related Educational Programs (CACREP, 2015) requires both the completion of an advanced degree program and passing an exam.

Science education licensure varies greatly between states. In some states each subject area within science (e.g., biology, chemistry, physics) is licensed with each area requiring specific content mastery and a test. Other states have comprehensive science licensure with which teachers can teach any area of science. Some states have licenses for grade levels. Middle school (generally 6th–8th grade) science is often a separate licensure from high school (9th–12th).

DEGREE REQUIREMENTS FOR SECONDARY SCIENCE TEACHERS

While universities vary in their requirements for science teacher preparation, three general benchmark levels can be found at the initial licensure level. The first benchmark is admission to teacher education and generally occurs in the second year of the program. Most universities require some basic skills exam, a minimal GPA, recommendations, and an interview or vetting process.

The second benchmark is prior to the final clinical field experience (i.e., internship or student teaching) and is the point at which most or all content courses have been completed. At this point in a candidate's program, they must have maintained a minimum GPA and a content exam has been passed to ensure competency in a subject matter. Other program-specific requirements may also be checked at this point to ensure the candidate is ready to begin the final field experience.

The final benchmark is at the end of the final clinical practice in which the candidate is required to demonstrate their ability to apply their knowledge and skills in a clinical setting to impact PK–12 student learning. Programs leading to advanced licensure have similar benchmarks appropriate to the licensure level, state requirements, and/or national accrediting requirements.

The vast majority of the benchmarking events above address accountability requirements from state and national agencies and/or accrediting bodies. CAEP's Standard 3 sets ambitious candidate recruitments and selectivity criteria including high GPA and an increasingly higher average performance on nationally normed assessment tests (CAEP, 2013). No such requirements exist for the majority of alternative certification programs. Significantly, a prospective science teacher who can not be admitted to a preparation program can, after graduation and passing the required exams, enter the classroom with the same credentials as a prospective science teacher who has met the requirements above yet has also met program requirements such as demonstration of the ability to work with all students, a knowledge of safety, an ability to impact P–12 student learning, and a strong content knowledge as demonstrated by a minimal GPA. Two science teachers, from two different pathways, can be found side-by-side in the same school. While there is no guarantee of the success for any beginning teacher, research indicates one will be successful, the other perhaps less so (Darling-Hammond et al., 2005).

CONTENT AND PEDAGOGY REQUIREMENTS FOR SECONDARY SCIENCE PREPARATION PROGRAMS

Bachelor degrees at the majority of U.S. institutions are limited by accreditation requirements in the maximum credit hours that can be required of any major. As a result, the majority of college degree programs require an average of 120 credit hours with some variation among programs and degrees. Science teacher preparation programs fall within the range of a credit hour requirement of greater than 120 hours but less than 128. (Johnson,

Reidy, Droll, & LeMon, 2012). Only programs in engineering had higher credit hour requirements than teacher education majors.

The typical teacher preparation program has on average 51 hours in general studies (liberal arts), 38 hours in their content, 28 hours of professional teacher education courses, and 14 clinical hours (Coleman, Coley, Phelps, & Wang, 2003) with great variety noted even for programs within the same state (Johnson et al., 2012).

As teacher education requirements moved from input measure to outcome, accreditation bodies no longer required specific courses. However, demands on the training of teachers, including mandates from states and federal laws, often require programs to offer subject-specific courses that become de facto course requirements. These often include courses covering educational technology, meeting the needs of students with disabilities, English for speakers of other languages, and reading across the curriculum. In addition, the majority of programs offer courses that cover educational assessment, adolescent psychology, classroom management, and instruction. These courses are general pedagogy courses and usually fall within the professional teacher education sequence. Offered predominately within a college of education, the courses are taught by education faculty and not faculty from specific content areas.

The NSTA Standards for Science Teacher Preparation (2012) do not require programs undergoing review to have specific content courses; however, for a program to be nationally recognized, there must be a 90% alignment between the NSTA Content Analysis Form (NSTA, 2012) and program coursework. The Content Analysis Form is a collection of scientific concepts based on discipline-specific recommendations of the American Association of Physics Teachers, the American Chemical Society, the National Association of Biology Teachers, and the National Earth Science Teachers Association and is aligned with *A Framework for K–12 Science Education: Practices, Crosscutting Concepts, and Core Ideas* (Council, 2012). The requirements are designed to prepare teachers with the content knowledge necessary to be able to effectively design and implement curriculum in classrooms that promote student learning in the sciences. Each preparation program determines the courses at their institution that meet the required content knowledge. The state-required content exam is also a measure of content knowledge with minimal passing scores required for licensure. For programs that are not required to undergo NSTA review, content courses are also not specified, but instead, each program must provide evidence of candidate performance on program-specific competencies as well as pass rates on state-required content exams for Title II reporting.

Teaching science so that all students learn requires more than just knowledge of content and basic pedagogy. Successful science teaching requires pedagogical content knowledge (Shulman, 1987). Pre-service science teachers' first exposure to how to teach science is usually found in a program's science-specific methodology course. While there are no requirements for such a course by any accrediting body, to be able to address standards that require graduates to demonstrate that they have the science-specific pedagogical knowledge to teach all students, most programs have at least one science methodology course. Depending on the size of the program, the science methods course may be subject specific, (i.e., a methods course for biology or chemistry) but much more common is a general methods course in which the activities a student does are focused on the field of science they will teach. Additionally, to further provide instruction within a specific field of science, a student may be placed in a field experience in a classroom within their area.

The instructors of science methodology courses can be faculty from either a science department or from an education department depending on where the program is located. Regardless of the location, program standards for science accreditation (local, state, or national) are primarily addressed by activities in the methodology courses. It is in these courses that beginning science teachers learn the knowledge and skills to produce lessons that address state or national standards for students. Such knowledge and skills are delineated in the NSTA Standards for Science Teacher Preparation.

FIELD EXPERIENCE REQUIREMENTS FOR SECONDARY SCIENCE PREPARATION

National standards have set specific expectations for both the quality and quantity of clinical field experiences and are indicative of state and national expectations for field experience requirements (CAEP, 2013). The standards require partnerships with local P–12 schools that are mutually beneficial and have clear entry and exit expectations. Both the program and the P–12 partners co-select, prepare, evaluate, and retain the clinical educators who work with the student teachers in their classrooms, and both must provide clinical experiences of sufficient depth and breadth to ensure all graduates have a positive impact on P–12 student learning. Performance-based assessments at key points within the program are required to demonstrate candidates' development. Of note is the lack of specificity for the number of hours for clinical placements; however, other requirements such as requiring placements in diverse settings result in multiple placements at different levels of the program. Field placements are traditionally done at least three times in a program. The first is usually done in an introductory course and is labeled as the early field placement. Time requirements vary from 15 to 30 hours in a classroom, and the placement is in the candidate's area of licensure. Expectations of early field experiences are observation of the clinical faculty, tutoring, and perhaps small micro-lessons.

The next level of field experience is usually found in a methods course and/or an upper level professional teaching course (i.e., adolescent psychology). Because the placement is later in the candidate's program, the expectations for the experience include greater interaction with the P–12 students, more responsibility for designing and/or delivering the curriculum, assessing students, and interaction with parents and other faculty.

The final field experience in a science preparation program is known as "student teaching." While there are no national requirements for the length of student teaching, most states set a minimal length of time for student teaching. For the 2009–2010 reporting year (USDOE, 2012) the median average number of hours required of supervised clinical experiences was 109 for all teacher preparation programs. The number of hours was higher from traditional programs (120) and much lower (60) for alternative programs based at institutions. For pre-service science teachers, this experience is critical in their development of the ability to design a curricular unit that demonstrates their understanding of content, how best to present that content, and an understanding of students' misconceptions of that content. The curricular unit that each candidate prepares is an artifact of their training. Candidates go through a pre-service science program and learn the tools they will need within the classroom to reach all students. These tools include content, pedagogy, and pedagogical content knowledge. It is in the student teaching experience that candidates have the opportunity to use the tools to build knowledge for students.

ESSENTIAL ELEMENTS OF SCIENCE PREPARATION PROGRAMS

Students in the United States are being asked to learn science in a very different way than was previously required. With the publication of the *A Framework for K–12 Science Education* (NRC, 2012) and the subsequent development of the NGSS came a clear delineation of what it means to be proficient in science. Using three dimensions (practices, crosscutting concepts, and disciplinary core ideas), the NGSS present a view of science significantly different from the existing science standards developed in the 1990s (NRC, 1996). The NGSS present not only the science content K–12 students should learn at what level, the standards include the way scientists and engineers engage in their work through the scientific and engineering practices. The crosscutting concepts reach across the disciplines and link them with concepts that provide a framework for students to build an understanding of natural phenomenon rather than learning isolated facts within a discipline. Rather than having a wide range of content to cover, the NGSS delve into core concepts in depth. This represents a radical change from how science is currently taught in classrooms in the United States and as such, requires significant changes in the way science teachers are prepared.

While the NGSS are K–12 standards, secondary science preparation programs must provide future teachers with a firm understanding of the theories and concepts that support the NGSS as well as the interrelationship of the three dimensions. If a teacher understands why the standards were developed the way they were, they will understand better how to develop curriculum to present the standards to students.

However, addressing the NGSS in science preparation programs is not enough. Science teachers must also gain critical knowledge about laboratory safety, teaching for all students, pedagogical content knowledge, classroom management, and professionalism. To assume the only thing a future science teacher needs to know is the NGSS is to ignore the realities of the classroom. Science preparations programs have the monumental task of giving future teachers the tools they need to construct learning and provide them classroom environments to practice building student knowledge in. It is a mistake to call them ready to teach after completing a science preparation program. To believe so is like teaching a contractor how to use a hammer and nails and expecting them to build a building their first time working with lumber. Science teacher preparation cannot end at graduation. Beginning science teacher induction matters (Wang, Odell, & Schwille, 2008).

It is far more worthwhile to conceive teaching as a continuum that originates in a science teacher preparation program and continues with all stakeholders working together to develop, not produce, teachers over time. Science teacher preparation program responsibilities do not end with graduation. There is evidence of some movement to view teaching this way. For example, CAEP Standard 2 requires significant mutually beneficial partnerships between higher education and school systems. In such a system, clinical practices are part of the core experience for future teachers, (NCATE, 2010) and K–12 schools benefit from partnerships with universities through professional development opportunities. Such a system must be mutually beneficial and designed together to ensure teachers and science teacher preparation programs own the process and design effective instruction for all students grounded in mutually beneficial professional development.

SUMMARY

There are many routes to science teacher licensure in the United States. Science teacher preparation programs are the traditional path to teaching and have a significant number of state and national standards and requirements that are measured as part of an accountability system. These requirements include evidence of both content and pedagogical knowledge and skills. Alternative licensure programs, in general, require a degree relevant to the area of licensure and a minimum of 24–28 semester hours in the major to be considered for licensure. As noted previously, pedagogical requirements for alternative licensure are minimal or non-existent.

As shortages increase and the pool of applicants decrease, state requirements can be relaxed and non-highly qualified persons certified under emergency licenses. For example, for the 2015–2016 school year the state of Oklahoma granted over 950 emergency certificates. The requirements for emergency certification include the documentation of need by a school superintendent, a college degree, and having passed, or be registered to take, the required subject area test (OSDE, 2015). This represents approximately 30,000 students without highly qualified teachers. Of those, approximately 15% were in the science areas. Emergency certification is valid for one school year eliminating a passing score on the subject area test as a true requirement for licensure. No requirements for pedagogy or content pedagogy are listed.

Science teachers who graduate from state and/or nationally approved preparation programs have met strict guidelines for content, pedagogy, and content pedagogy. They have demonstrated knowledge of what to teach, how to teach it, and how to ensure student safety in laboratory settings. Emergency or alternative licensure teachers have not necessarily met these same high standards. As we demand more and more of traditional science teacher preparation there is an inherent inequity between licensure routes. Yet, in the end, regardless of how a science teacher gets into the classroom, they must address student learning. While certainly there are examples of excellent teachers from all routes, in the end, the expectations for all science teachers remain the same: Teach the students that show up and teach them well.

REFERENCES

American Association of Colleges for Teacher Education. (2012). *edTPA*. Retrieved from http://edtpa.aacte.org

Coleman, A. B., Coley, R. J., Phelps, R. P., & Wang, A. H. (2003). *Preparing teachers around the world.* Princeton, NJ: Educational Testing Service.

Committee on Prospering in the Global Economy of the 21st Century: An Agenda for American Science and Technology, National Academy of Sciences, National Academy of Engineering, and Institute of Medicine (U.S.). (2007). *Rising above the gathering storm: Energizing and employing America for a brighter economic future.* Washington, DC: National Academies Press.

Council for Accreditation of Counseling and Related Educational Programs. (2015). Understanding accreditation. Retrieved from http://www.cacrep.org/value-of-accreditation/understanding-accreditation/

Council for the Accreditation of Educator Preparation. (2013). The CAEP Standards. Retrieved from http://caepnet.org/standards/introduction

Council of Chief State School Officers. (2011). *InTASC model core teaching standards.* Retrieved from http://www.ccsso.org/Resources/Resources_Listing.html?search=model+core+teaching+Standards

Darling-Hammond, L., & Rothman, R. (2015). *Teaching in the flat world: Learning from high-performing systems.* New York, NY: Teachers College Press.

Darling-Hammond, L., Holtzman, D. J., Gatlin, S. J., & Heilig, J. V. (2005). Does teacher preparation matter? Evidence about teacher certification, teach for America, and teacher effectiveness. *Education Policy Analysis Archives, 13*(42), n42.ce education. *Handbook of Research on Teacher Education,* 459–484.

Educational Testing Service. (2015). Praxis series. Retrieved from https://www.ets.org/praxis/about?WT.ac=praxishome_about_121126

Gardner, D. P. (1983). *A nation at risk.* Washington, DC: The National Commission on Excellence in Education. United States Department of Education.

Girod, G. R. (2002). Connecting teaching and learning: A handbook for teacher educators on teacher work sample methodology. Washington, DC: AACTE.

Heilig, J.V. & Jez, S.J. (2010). *Teach for America: A review of the evidence.* Boulder, CO: Education and the Public Interest Center & Education Policy Research Unit. Retrieved from http://epicpolicy.org/publication/teach-for-america

Humphrey, D., & Wechsler, M. (2007). Insights into alternative certification: Initial findings from a national study. *The Teachers College Record, 109*(3), 483–530.

Johnson, N., Reidy, L., Droll, M., & LeMon, R. E. (2012). Program requirements for associate's and bachelor's degrees: A national survey. In *Technical Report.* HCM Strategists.

Lackney, J. A. (2015). History of the schoolhouse in the USA. In *Schools for the Future* (pp. 23–40), New York, NY: Springer.

Marzano, R. J. (2012). The two purposes of teacher evaluation. *Educational Leadership, 70*(3), 14–19.

Minow, M. (2015). The common school before and after Brown: Democracy, equality, and the productivity agenda. *Yale Law Journal,* 120:1454.

National Center for Education Statistics. (2014). *Teacher attrition and mobility: Results from the 2012–13 Teacher Follow-up Survey* (NCES 2014-077). Washington, DC: United States Department of Education.

National Council for Accreditation of Teacher Education. (2015). *About NCATE.* Retrieved http://ncate.org/Public/AboutNCATE/tabid/179/Default.aspx

National Council for Accreditation of Teacher Education. (2010). *Transforming teacher education through clinical practice: A national strategy to prepare effective teachers.* Washington, DC: Author.

National Council for Accreditation of Teacher Education. (2008). *Professional standards for the accreditation of teacher preparation institutions.* Washington, DC: Author.

National Council on Teacher Quality. (2014). *Teacher prep review 2014.* Retrieved from http://www.nctq.org/teacherPrep/review2014.do

National Governors Association Center for Best Practices & Council of Chief State School Officers. (2010). *Common Core State Standards.* Washington, DC: Authors.

National Research Council. (2012). *A Framework for K–12 Science Education: Practices, Crosscutting Concepts, and Core Ideas.* Washington, DC: The National Academies Press.

National Research Council. (Washington). Division of Behavioral and Social Sciences and Education. Committee on the Study of Teacher Preparation Programs in the United States. Center for Education. (2010). *Preparing teachers: Building evidence for sound policy.* Washington, DC: National Academies Press.

National Research Council. (1996). *National Science Education Standards.* Washington, DC: The National Academies Press. doi:10.17226/4962

National Science Foundation. (2015). *Robert Noyce teacher scholarships.* Retrieved from http://www.nsf.gov/pubs/2015/nsf15530/nsf15530.htm

National Science Teachers Association. (2012). *Standards for science teacher preparation.* Retrieved from http://www.nsta.org/preservice/

NGSS Lead States. (2013). *Next generation science standards: For states, by states.* Washington, DC: The National Academies Press.

No Child Left Behind. (2002). Act of 2001, Pub. L. No. 107–110, § 115. *Stat, 1425,* 107–110.

Oklahoma State Department of Education. (2015). *Emergency certification.* Retrieved from http://www.ok.gov/sde/emergency-certification-administrator-use-only

Panel, B. R. (2010). *Transforming teacher education through clinical practice: A national strategy to prepare effective teachers.* Washington, DC: National Council for Accreditation of Teacher Education.

Presley, J. B., & Coble, C. R. (2012). *Seeking consensus on the essential attributes of quality mathematics and science teacher preparation programs.* APLU/SMTI, Paper 6. Washington, DC: Association of Public and Land-grant Universities. Retrieved from http://www.aplu.org/document.doc?id=4098.

Shulman, L. (1987). Knowledge and teaching: Foundations of the new reform. *Harvard educational review, 57*(1), 1–23.

Teach for America. (2015). *Annual Report.* Retrieved from https://www.teachforamerica.org/annual-report

Troops to Teachers. (2015). *Eligibility requirements.* Retrieved from http://troopstoteachers.net/AbouttheProgram/Eligibility.aspx

The National Association for Alternative Certification. (2015). *External talking points.* Retrieved from http://www.alternativecertification.org/wp-content/uploads/2015/11/External_Talking_Points_NAAC_final.pdf

United States Department of Education. (2014). *Title II reports on the quality of teacher preparation and states' requirements and assessments for initial credentials.* Retrieved from http://www2. ed.gov/about/reports/annual/teachprep/index.html

United States Department of Education. (2015a). *Distribution of teacher education assistance for college and higher education (TEACH) grant program funds by institution.* Retrieved from http://www2.ed.gov/finaid/prof/resources/data/teach-institution.html

United States Department of Education. (2015b). *About Ed/Overview.* Retrieved from http://www2.ed.gov/about/overview/fed/role.html

United States Department of Education. (2015c). *Teacher shortage areas.* Retrieved from http://www2.ed.gov/about/offices/list/ope/pol/tsa.doc

U.S. Department of Education, Office of Postsecondary Education. (2012). Higher Education Act Title II Reporting System. Washington, DC: Author.

Veal, W. R., & Allan, E. (2014). Understanding the 2012 NSTA science standards for teacher preparation. *Journal of Science Teacher Education, 25*(5), 567–580.

Wang, J., Odell, S. J., & Schwille, S. A. (2008). Effects of Teacher Induction on Beginning Teachers' Teaching A Critical Review of the Literature. *Journal of Teacher Education, 59*(2), 132–152.

CHAPTER 10

SCIENCE TEACHER EDUCATION IN ARGENTINA

Melina Furman
Universidad de San Andrés

Mariana Luzuriaga
Universidad de San Andrés

In Argentina, as in many other countries of the world, science education has been declared a priority, a degree that acknowledges the role of scientific literacy as a fundamental part of the education of future citizens. Yet, low student performances on national and international science assessments, as well as high rates of secondary student drop-out and measurable inequality amongst schools have shown that there is a profound (and urgent) need of improvement in this terrain. It is precisely within this context that the debate on how to prepare science teachers takes on deeper meaning; that is, by recognizing the key role these future educators play on influencing and fostering student achievement.

This chapter presents an analysis of Argentina's teacher education system and examines its preparation of future secondary science teachers. In order to frame the analysis within the local context, the text begins by briefly describing the structure and current challenges of the educational

Model Science Teacher Preparation Programs, pages 205–227

system in Argentina. Then, it examines in greater depth the teacher education system, focusing on the preparation of science teachers, underlining the system's strengths while raising questions about the challenges for future improvement.

A BRIEF LOOK AT THE ARGENTINE EDUCATIONAL SYSTEM

The Argentine education system is regulated by the National Law of Education, which was enacted in 2006 to abrogate the educational reform initiated in 1993 under the previous Federal Law of Education. Some of the major changes introduced with the 2006 reform were the reorganization of the education system's structure, the extension of compulsory education from 5 years of age to the end of the secondary level, and overall the reform of teacher education. In addition, the law stipulates that the national budget for education must not be below 6% of the country's gross domestic product (GDP; National Law of Education, 2006).

The system consists of four levels: preschool, elementary, secondary, and higher education. Children attend preschool from 2 to 5 years of age, with the last two years (K4 and K5) being compulsory. Elementary education is intended for children 6 years of age and older and can last between 6 and 7 years depending on the jurisdiction of the school. Secondary education lasts between 5 and 6 years, resulting in 14 years of compulsory education in total.

Working conditions for teachers in Argentina are very heterogeneous. Factors including assigned workload and teacher salaries vary substantially between provinces. Student–teacher ratios also vary within the country, particularly in rural areas. On average, secondary classes have 28 students (Organization for Economic Development and Cooperation [OECD], 2013).

Regarding workload, at the secondary level in particular, 45.7 % of teachers work 24 hours or less per week, whereas 31.8 % of teachers work between 25 and 48 hours, and 10.4 % work 49 hours or more (DiNIECE, 2006). It is highly common for secondary teachers to occupy positions in several schools. It is also worth noting that all of the hours reported above indicate the number of hours spent in front of class. In general, there is no extra paid time for teachers to work on planning, grading student work, attending staff meetings, or engaging in professional development. This lack of paid time for other tasks other than teaching to students is, undoubtedly, one of the factors that most severely impacts the chances of improving teaching quality in the country.

Teacher salaries are relatively low, in comparison to other professions. Using a teacher with 10 years of experience as a reference point, in December of 2013, yearly salaries at the secondary level varied between US $5,328 and US $10,536 for a 15-hour weekly workload in the same period, depending on the province (CGECSE, 2014). Salary differences (i.e., additions to the base salary) are based solely on seniority, as well as other factors such as school location. For instance, teachers working in disadvantaged areas are often eligible for extra income.

Concerning the school curriculum, Argentina currently holds a set of national standards (called Priority Learning Guidelines), which direct the development of curriculum guidelines within each state (Federal Board of Education, 2004). In science, the national standards endorse scientific literacy as a key learning goal for all students and support inquiry-based pedagogies, such as engaging students in guided investigations of natural phenomena, analyzing data and participating in debates over scientific issues (Ministry of Education, 2007). Based on national standards, each province designs its specific curriculum.

Improving both education quality and equity is currently an urgent issue in the country. Argentine students have performed poorly on recent national and international examinations, both in science and in other subject areas (DiNIECE, 2010; OECD, 2013; Sequeira, 2009). These examinations have also shown significant differences in student performance according to their socioeconomic status, reflecting a profoundly inequitable educational system. For instance, Argentina was ranked 59th out of the 65 countries participating in the Programme for International Student Assessment (PISA) 2012. Argentina's mean score in science was 406, which was well below the OECD average of 501. Almost 60% of Argentine students showed a science proficiency level of 1 or below (approximately 30% of students were at level 1, while 30% were below level 1).

On the positive side, Argentina has experienced a remarkable expansion in enrollment rates at the secondary level during the last decades (OECD, 2014a). Secondary education coverage improved significantly, increasing at a rate of roughly 10% in the last decade (Guadagni Alieto, Lasanta, & Álvarez, 2014). Currently, about 80% of young people aged between 12 and 17 years attend school at the secondary level. However, despite these increasing numbers, less than 50% of the student population graduates secondary level in a timely manner, (Guadagni Alieto et al., 2014) and there is a persistent inequality amongst provinces in terms of school retention and completion rates. For instance, the annual rate of student drop out at this level reaches 15%, and the repetition rate exceeds 6% (Guadagni Alieto et al., 2014). This deficiency is also reflected in the high levels of youth unemployment, which in 2014 affected 17.7% of the population aged between 18 and 24 (Instituto Nacional de Estadística y Censos [INDEC], 2014). Consequently, various efforts are being undertaken regarding the need to support student learning trajectories and prevent student dropout, with different levels of impact (Jacinto, 2009; Tenti Fanfani, 2009). Initiatives include providing schools with tutors that oversee student learning in all subject-matter areas or offering extracurricular pedagogical support for students in the particular subject areas that they are failing in.

A FRAGMENTED TEACHER EDUCATION SYSTEM

In order to understand the characteristics of teacher education in Argentina, both for science and other subject matters, it is important to have a brief look at the general structure of the system and its recent reform process. In this section of our analysis, we will focus on the preparation of pre-service teachers.

Since its origins, secondary teacher education in this country depends on two differentiated institutional and organizational contexts: higher education teacher education institutes (Institutos Superiores de Formación Docente, in Spanish), which are tertiary institutions (i.e., not university-based), and universities, both managed by the public or private sector.

Initially, in a context in which secondary school was conceived as a more exclusive education track that prepared leading sectors for university, it became "natural" for secondary teachers to be university graduates as well (Perlo, 1998). However, in 1904 the first national teacher education institution for secondary teaching was created in Buenos Aires, giving origin to the binary character of teaching degrees as either tertiary education or university.

One of the distinctive characteristics of the latter is that teacher educators hold a university degree themselves, whereas this is not a mandatory requirement in tertiary education. On the other hand, university teacher education degrees typically offer a deeper and more comprehensive disciplinary preparation and demand students to engage in scientific and educational research experiences, which is one of the major pending challenges in teacher education institutions (regardless recent initiatives to enhance research as part of their pedagogical offer, as we will refer to later) (Ministry of Education, Science, and Technology, 2012).

Such conditions result in distinct educational offers, further accentuated by the traditionally lacking interaction amongst both subsystems, whether through support, exchange or collaboration experiences (Federal Board of Education, 2007b). The coexistence of both subsystems led to a great diversity in terms of the length of studies, curricular contents and degree accreditation (Davini & Alliaud, 1995), which still persists at present despite the renovated efforts to unify and standardize teacher education programs throughout the country with the 2006 reform.

The system's fragmentation is key to understanding some of the challenges teacher education presents in Argentina. To this regard, we join others who claim the need to prioritize and strengthen teacher training by gradually making it university-based, as it is in many other countries. However, we also recognize that this is a not an easy or straight forward goal to attain, but demands engaging in profound debates on the purposes of teacher education and a strong commitment to pursue both realistic and effective reform policies.

In the first place, this is a challenge that would imply a profound structural change, for traditionally higher education teacher education institutes had a predominant role in teacher education. This tendency still continues nowadays: only 17.9% of the practicing secondary teachers hold a university teaching degree (Federal Board of Education, 2007f).

We also think that important questions should be addressed about the transformative potential of moving the teacher education system to a university context and under which conditions. In Vaillant's (2013) words, "even if teacher education is successfully emplaced at university level, this does not guarantee that the quality of teacher education will be improved *per se*" (p. 199). In this sense, there is the risk of prioritizing a formal structural change without revisiting the most substantial aspects of teacher education. On the contrary, we agree that improving the quality of teaching and strengthening professional skills (whether at higher education institutes or universities) demands revising key aspects such as the teacher education programs' contents and pedagogical approaches, enhancing the integration of theory and practice, and reinforcing the articulation with the schools (Perrenoud, 2004). Therefore, the aim should not be the consolidation of the system as such, but how to improve in the preparation of teachers in the country.

Recent efforts have been made to make teacher education more coherent. Without questioning the system's general structure, but in an attempt to unify and standardize teacher education programs throughout the country through a renovated reform process, in 2006 the National Law of Education Nº 26.206 and the creation of the National Institute of Teacher Education (INFD, for its Spanish acronym), were enacted. The INFD's main goal was to develop initiatives that could combine jurisdictional particularities with a shared national vision to re-establish teacher education as a strategic priority for the improvement of education as a whole.

We believe a positive aspect of this process is the commitment towards the consolidation of the teacher education system in Argentina, unified in terms of training quality and with nationwide validation for every program. However, this overhaul also requires the availability of significant resources and the creation and implementation of necessary support systems, which are not always available, especially in the most disadvantaged provinces of the country.

Since 2007, teacher education programs and institutions are to be adjusted to national requirements. Taking national guidelines as a general framework, each province has to elaborate their own curricular documents for each teaching degrees and have it approved by the INFD. Yet, bureaucratic obstacles have emerged, and there are still many programs that have not been assessed or permanently accredited; this on-going process has been fulfilled slower than expected and with different levels of compliance (Cámpoli, 2004).

The recent reform process did not only involve structural changes, but also the revision of teacher education curricula and contents. The Federal Board of Education particularly highlighted the need to evaluate and update teacher education curricula and programs as a key initiative to reinforce knowledge and teaching strategies related to literacy, math and science (Federal Board of Education, 2011c).

These changes respond to a renewed interest in adapting the educational system (and teacher education in particular) to the demands of contemporary knowledge societies, technological advances and socio-political and economic circumstances (Esteve, 2006). The founding principles of education systems, linked to the construction of national identity and citizenship, gave way to an economic thrive which focused on educating competent human resources and a social dimension, related to inclusion and equity goals (Tedesco, 2007).

WHAT IS REQUIRED TO BE A SCIENCE TEACHER IN ARGENTINA?

In Argentina, there is no formal selection of teacher candidates. Applicants for teacher education programs for both the elementary and secondary levels are not required, neither at universities nor at higher education teacher education institutes, to take academic exams prior to admission. The only conditions are that program candidates must have a secondary degree certification (Law of Higher Education, 1995, art. 7) and, in some jurisdictions, must pass a psychophysical health examination (as in many other degrees). In addition, some jurisdictions or particular institutions may also require their students to attend an introductory course, which is often part of their first year of study.

It is not clear how this lack of early selection impacts the quality of the country's teachers. It is worth noting, however, that this policy is framed under the tradition of open admission to most higher education institutions (including universities) in the country.

On a positive note, this may constitute an opportunity, given that recruiting teacher candidates is actually a major challenge of the teacher education system in the country, which is facing a qualified teacher shortage in many areas, especially in science and mathematics for the secondary level.

Various reasons account for this difficulty, such as the diminished social and material status of the teaching profession (Mezzadra & Veleda, 2014; Ministry of Education, 2008; Tenti Fanfani, 2005; UNESCO, 2012).

Secondary teaching degrees have a minimum workload of 2.600 hours in a total of four years (National Law of Education, 2006, art. 75; Federal Board of Education, 2007d; art. 26). In science, there are currently four main degrees that teachers can obtain: biology, chemistry, physics or natural sciences.

Every teacher education program is organized into three different fields of knowledge: general pedagogical education, specific education (or content pedagogical courses) and professional exercise (or specific field experience; Federal Board of Education, 2007d, art. 30). Between 25 and 30% of students' total workload must focus on general education, between 50 and 60% must focus on specific education, and between 15 and 25% must focus on professional exercise.

At the end of their degree, prospective teachers do not have to take any national or state examinations, nor produce a thesis, as a requirement for acquiring their license. Again, further research is needed to assess how this lack of national assessment impacts on the quality of the country's teachers, although some argue it is indispensable to improve teacher education (Mezzadra & Veleda, 2014; UNESCO, 2012).

This absence of any national assessment can be understood within the context of decades of system decentralization initiated with the 1990s reform, which transferred the authority over the educational system from the national level to each jurisdiction, consequently adding to the system's fragmentation described above.

Also, it is important to take into account that teacher evaluation has traditionally been a particularly problematic issue in Argentina, as it is deeply rejected by teachers and teacher unions who consider it a threat linked to a market-based conception of education (Perazza & Terigi, 2008; Tenti Fanfani, 2005). At present, only a few provinces implement voluntary teacher assessments for in-service teachers, as an exception rather than the norm (see, for example, the Teacher Assessment Program from the city of Buenos Aires).

Lately, a number of initiatives have been put into place, designed to incorporate national or provincial teacher assessments in teacher training. One, the National Plan for Teacher Education 2012–2015 mentions the need to establish a comprehensive assessment of the teacher education system, in order to identify its strengths and areas for further improvement. For this purpose, it commends the incorporation of annual assessments for students beginning in the second year of their studies. It is worth underlining that these assessments do not intend to evaluate students, since they do not have direct consequences on their chances of graduation; rather, they assess the educational institutions themselves (Federal Board of Education, 2012b). On a similar note, a draft bill has been recently presented at the

Congress to assess student teachers at a national level (Draft Bill 1691/15), but has not been approved yet.

Such initiatives, in spite of not being materialized in practice yet and being the subject of current political debate, show revamping of licensure requirements is an issue on the agenda. However, in order to implement these or other initiatives on teacher assessment, both in the pre-service and the in-service levels, there is still a huge need to establish political consensus on the matter of teacher assessment, which does not seem an easy goal in the short term (Perazza & Terigi, 2008).

THE SCIENCE TEACHER EDUCATION CURRICULUM

Regarding the curriculum, the preparation of science teachers in Argentina is framed under a more comprehensive philosophy of teacher education, which underlines teachers' roles in the construction of a just and democratic society for all. As such, national curricular guidelines for teacher education define teaching as a

> [r]eflexive and critical practice of cultural mediation, characterized by the ability to contextualize teaching interventions that promote student learning and support democratic processes within educative institutions, driven by ideals of justice that guarantee better and more dignified living conditions for all students. (Federal Board of Education, 2007d, p. 25)

Based on these ideals, teacher education programs promote an integral approach combining subject-matter knowledge and skills, pedagogy and interdisciplinary strategies. In order to clarify the general structure of the curriculum, an example of the complete program for secondary school biology teachers for one of the largest provinces of the country is presented in the Appendix. Next, we analyze in more detail the different phases of the curriculum: general pedagogical education, specific education, and professional exercise, which we describe in more detail below.

General Pedagogical Education Courses

General pedagogical education refers to the domain of conceptual and interpretative frameworks on general issues related to education, teaching and learning. It is

> meant to develop a solid humanistic education; the domain of conceptual and interpretative frameworks and value for the analysis and understanding of culture, time and historical context, education, teaching and learning; and

> the development of professional judgement to act in different socio-cultural contexts. (Federal Board of Education, 2007d, art. 30.1)

National curricular guidelines for teacher education emphasize that the main aim of teaching is to promote student learning in real and diverse contexts. This goal implies that teachers have to position themselves as reflective professionals, developing skills "to analyse their everyday practice and contribute to its continuous improvement, taking into consideration its particular context at social, institutional and classroom levels" (Federal Board of Education, 2007d, p. 26). Such an analytical and reflective capacity, alongside with other skills and dispositions aimed at offering equal and quality learning opportunities for all students, should be fostered during teacher training, as it is listed in Table 10.1.

It becomes evident, then, that teacher training is not only defined as requiring proficiency in subject-matter content, but that this should be complemented and enriched with other key skill sets to perform successfully as secondary teachers: knowledge on different pedagogical approaches,

TABLE 10.1 Abilities Involved in Teaching According to the National Curricular Frameworks for Teacher Education in Argentina

Teachers should be able to
a. Dominate content and update their theoretical frameworks.
b. Adapt, produce, and assess curricular content.
c. Recognize the educational purposes involved in teaching contents.
d. Broaden their own cultural horizon beyond the specific contents they have to teach.
e. Identify the students' learning characteristics and needs as the baseline for their teaching practices.
f. Organize and conduct learning situations taking into account the socio-political, cultural, and linguistic contexts they take place in.
g. Develop pedagogical approaches that take into account student diversity based on the firm belief that all students have the ability to learn.
h. Actively involve students in their learning process.
i. Support student learning by identifying their strengths and areas of improvement.
j. Manage time and create a positive classroom environment to enable all students to learn.
k. Enable learning at both collective and individual levels.
l. Make an efficient use of the available resources for teaching.
m. Select and make an appropriate use of ICTs according to the teaching context and purposes.
n. Recognize the contexts' characteristics and needs at school, family, and community levels.
o. Interact with the students' families to foster their involvement in the educative process.
p. Teamwork with other teachers, develop shared institutional projects and participate in school initiatives.

Source: Compilation based on the National Curricular Guidelines for Teacher Education (Federal Board of Education, 2007d).

teaching strategies, national history, political and legal organization, epistemological perspectives; and skills to work as a member of a team, develop student assessment, incorporate ICTs in teaching, amongst many others (Hisse, 2014). All these topics are covered within the general education courses of the program, and shared by all future teachers independently of their subject matter.

One of the main challenges regarding the field of general pedagogical education is that it is often boarded superficially, presented as a mere framework decontextualized from its epistemological grounds, instead of fostering a deep and complex understanding of the educative process (Federal Board of Education, 2007d). Yet, as we will discuss below, it is argued that programs give much emphasis to general pedagogy, with less instruction time devoted to specific preparation for science teaching, which limits teachers' preparation (Cofré et al., 2015).

Specific Education Courses

The specific education phase of teacher education programs are meant to address the analysis, formulation, and development of knowledge and teaching strategies for the particular level and subject-matter student teachers aim to graduate in (Federal Board of Education, 2007d). It is oriented to build an understanding of the particularities of the subject matter (biology, physics or chemistry, accordingly) teaching at secondary level, as well as its purposes in the education system and society as a whole (Hisse, 2014).

Therefore, specific education involves the study of the particular subject-matter content, pedagogical content knowledge (Shulman, 1986), students' particularities related to that subject and the main issues related to the given level or modality (Ferrata, 2014). Teacher candidates should not only demonstrate proficiency to dominate the essential subject-matter concepts and procedures, but to implement appropriate teaching strategies to enable learning for each content's, students' and contexts' characteristics (Hisse, 2014).

National guidelines state three irreducible goals for science learning: *know* science (as a process and as a product), *do* science and *communicate* science (Pogré, 2010, p. 11). An understanding of science as a social construction and human endeavor is promoted, while a critical look at the nature of science is stated, "questioning distorted ideas on its nature based on inductive, positivist, and empiricist epistemologies" (Hisse, 2014, p. 15). It is worth noting, however, that neither inquiry-based pedagogies nor the concept of science practices are explicitly mentioned, as it is the case with other countries' teacher education programs (see for example NGSS, 2013).

To this regard, guidelines foster an integrated and global approach of science, which combines the understanding of science as a product (a body of knowledge) and as a process (the skills and procedures through which this knowledge is produced). Particularly, several jurisdictional curricular guidelines propose "a historic and epistemological perspective through the incorporation of two specific units that problematize the conception of science and the processes of scientific knowledge production, considering its socio-cultural, historic, ethical and political dimensions" (Hisse, 2014, p. 18). Within each program, 3–4 courses are designated to the history and philosophy of science.

In addition, these guidelines emphasize the need to include experimentation as a key practice in teacher training, for it is one of the most characteristic features of scientific methodology. It is argued that teachers should implement such an approach in their own science classes for it fosters student motivation and learning: "[I]t is considered that having students design and perform their own experiments at school enables a better understanding of how scientists work to solve problems, which results in a deeper comprehension and reflection on the nature of science and its phenomena" (Hisse, 2009, p. 111).

Table 10.2 enumerates the main knowledge and skills science teachers are expected to dominate regarding science practices.

However, research has shown that this proposed vision of science, which integrates epistemological and historical aspects, as well as the development of science practices as a learning goal for future teachers, still presents profound challenges for teacher educators, whom in many cases have not been prepared themselves in such vision or practices (Cofré et al., 2015). Very few teacher educators have any personal experience in scientific or educational research, which also hinders their ability to engage with innovative teaching strategies or introduce students to more realistic perspectives on the nature of science and scientific practices (Adúriz-Bravo, 2009).

This is particularly problematic, since science education reforms worldwide suggest that students' understanding of the nature of science is a main educational outcome (Lederman, Antink, & Bartos, 2014), which, in turn, implies that understanding the nature of science should be a key aspect in teacher education.

Regarding subject-matter content in particular, curricular guidelines present quite an extensive program. National guidelines set it between 50 and 60% of the workload, although in some provinces it is higher (Federal Board of Education, 2007d). This is the field in which the programs for each particular subject-matter differentiates itself, for subjects acquire higher specificity. For instance, the program for biology teachers presented in the Appendix includes 17 subject-matter courses. Four of these courses include basic knowledge in mathematics, physics and chemistry, with the

TABLE 10.2 New Knowledge for Science Teachers According to the National Curricular Frameworks

Science teachers should know:

a. The methods of scientific production in the past and in the present.
b. The main notions of the core scientific theories.
c. Science teachers should be able to:
d. Recognize, propose, and find possible solutions for scientific problems.
e. Hypothesize.
f. Recognize and operate multiple variables.
g. Design simple experiments to test hypothesis.
h. Propose fieldtrips to test hypothesis.
i. Use investigation as a teaching strategy and differentiate it from scientific investigation.
j. Use scientific language and foster students to apprehend specific terms.
k. Reflect on their conceptions on science in order to promote student learning.
l. Select content and elaborate pedagogical proposals to foster significant learning, in accordance to curricular frameworks.
m. Value and work with particular emphasis on transversal, integrated social and natural science matters such as environmental education, health education and sexual education.
n. Stimulate student motivation through appropriate teaching methods and by fostering the exchange of ideas, creativity, debates, etc.
o. Contextualize scientific problems according to the classroom and students' characteristics.
p. Write reports.
q. Link scientific issues with the students' interest and everyday life situations.
r. Plan activities that enable students to understand key scientific concepts.
s. Use pertinent and varied teaching resources (ICTs, books, etc.).
t. Model the performance of science activities in terms of security, scientific rigor and precision.

Source: Compiled based on Hisse, 2009.

goal of providing teachers with a broader understanding of science. The other 13 are specifically focused on biology: general biology, cellular and molecular biology, animal biology I and II, plant biology I and II, ecology, biology of microorganisms and fungi, genetics and biotechnology, human biology, history of life on earth and evolution, environmental education, human biology and health.

Guidelines also promote the integration of different subjects (Hisse, 2014), with the aim of providing teachers with an interdisciplinary approach to science learning. Courses' descriptions do actually mention the need to establish relationships between them. For example, one of the teaching recommendations for "cellular and molecular biology" is to "articulate knowledge with 'basic chemistry,' 'organic and biological chemistry, 'basic biology, 'mathematical and physics models' workshops' and specific biology contents from second and third years" (Hisse, 2014, p. 43). Yet, it is not clear for teacher educators how to accomplish that interdisciplinary integration, although guidelines propose "integration workshops" as special courses where teacher educators can foster that goal.

Finally, pedagogical content knowledge is developed over two or three specific courses, as prescribed by national guidelines (interestingly, though, the term "pedagogical content knowledge" as proposed by Shulman (1986) is not mentioned as such within curricular guidelines, but referred as "special pedagogies"). For instance, the curriculum presented in the Appendix includes two courses: Science Pedagogy and Science Pedagogy: Biology.

As Ferrata (2014) points out, in most programs, there is a strong prevalence of subject-matter content subjects over the development of pedagogical content knowledge. Research has shown that, in most teacher education programs in the country, science methods courses are insufficient, specific pedagogical discussions focused on teaching content are scarce, and there are few formal opportunities to reflect on teaching practices and tools (Adúriz-Bravo, 2009). In addition, some science teacher educators demonstrated deficiencies in the domain of subject-matter content (Adúriz-Bravo, 2009).

Professional Exercise

Along with general education and specific education, professional exercise is meant to progressively complete teacher training. In this sense, this field of knowledge is proposed as the backbone of teacher education, integrating knowledge contributions from the other fields to analyse, reflect and experiment teaching in different social and institutional contexts (Federal Board of Education, 2007d).

Historically, this was usually considered one of the fields with greater difficulties in the preparation of teachers. For one, field work and professional practice was almost entirely conducted at the end of the program, under vertical approaches which envisioned "practice" as a direct application of "theory" (Federal Board of Education, 2007d). Secondly, there was no specific definition on the content courses at teacher education institutions must address for this phase, only generically described as "workshops" or "seminars." Consequently, one of the key aspects of the curricular reform initiated in 2007 to improve teacher education was to reinforce professional exercise, including experimentation, reflection and innovation as key aspects in this field of knowledge. (Federal Board of Education, 2012b).

To this regard, one of the aspects curricular guidelines particularly emphasize is the need to integrate professional exercise with the rest of the fields from the beginning of the program. It is considered of prior importance for teacher students to enrich theory and practice mutually:

> Professional Practice requires a multi subject-matter approach that combines General and Specific Education contributions, permanently articulating

> theory and practice. The recommended contents for each year resume and re-elaborate on concepts developed in the other fields through a non-applicationist approach. (Hisse, 2014, p. 20)

Jurisdictional guidelines establish one professional practice subject per year along the whole degree. The very designation of these subjects (professional practice I, II, III, and IV) suggests certain level of continuity between them. Moreover, when analysing the particular descriptions of each, there are explicit references to contents from other fields of knowledge. For example, it is suggested that "professional practice I" integrates contributions from "educational ethnography" and the use of audio-visual resources, contents addressed by general education in the first year of training (see the Appendix).

On the other hand, it is clearly stipulated that the total workload for this field of knowledge should be distributed between teacher education institutes and associated secondary schools, where student teachers can perform projects, practice short lessons, obtain helper or tutor positions, etc. However, at this level of curriculum development, the correspondent workload for each learning emplacement is not clearly established. To this regard, we believe that further efforts are needed to establish stronger, more systematic and bidirectional bonds between teacher education institutes or universities and schools. Particularly, the role of schools and practicing teachers in the training process of future professors should be enhanced during the teaching placement phase and through other prior formative experiences. Furthermore, this should not exclusively depend on the possibilities or willingness of each particular institution, but be fostered and supported by higher levels of management.

Finally, regarding this field's contents, the importance of developing reflective capacities in student teachers is particularly highlighted. Both national and jurisdictional guidelines define this as a key element for teacher training and professional development. Curricular guidelines emphasize that the ability to reflect on one's own practice and analyse others' is essential for teaching, and that teacher education programs should foster it through different practical experiences (including, but not limited to, working on experiences, memories, evidence, narrative accounts, autobiographies, classroom observations, fieldwork, etc.).

Despite recent efforts, it has been pointed out that, still, professional exercise is often dissociated from the other fields of knowledge, with the articulation of theory and practice remaining a major challenge for many institutions (Ferrata, 2014). As in many other countries, this is a defying area for teacher educators, and further efforts still need to be done to help student teachers question their traditional epistemological perspectives on science learning and teaching to enrich their own practice (Loughran,

2014). As it has been already mentioned, this challenge implies establishing a closer link between teacher education institutions and schools. In this sense, authors point out that educative institutions and education in general improves when these two entities become centers for pedagogical innovation (Aguerrondo & Pogré, 2001).

BEYOND REFORM

Undoubtedly, the current attempt to overcome the system's fragmentation through the reform process and the creation of the INFD has been of profound importance for the preparation of teachers in Argentina. During the last years, degrees were unified nationwide and national and some jurisdictional teacher education curricular guidelines were successfully revised and updated to meet current standards. At present, teacher education national guidelines are consistent with the results of educational research and with international consensus on the best approaches and practices in science teaching (Pogré, 2010).

Moreover, under the frame of the "national strategy of improvement in science and math education" enacted by the Federal Board of Education, higher education institutes were equipped with information, multimedia and bibliographical resources, and various programs to foster professional development and postgraduate degrees were set into place (Federal Board of Education, 2011b). These initiatives contribute to comply with the basic necessary conditions for quality teacher education, but are not solely sufficient.

There are important challenges that still need to be considered to improve science education in the country. One of the biggest challenges is, perhaps, the inadequacy in teaching strategies and content that still persists both in schools and in teacher education (Adúriz-Bravo, 2009; Ministry of Education, 2007), which promote an encyclopaedic view of knowledge in general and science in particular. To this extent, although curricular guidelines are consistent with globally valued contemporary approaches to science teaching, there is a clear discordance between what is prescribed and what teachers actually put into practice (Furman & Podestá, 2009; Pasmanik & Cerón, 2005). Despite the rhetoric of constructivist and inquiry-based approaches, when analysing science lessons in both elementary and secondary schools:

> [W]e see that theoretical explanations and definitions prevail over experiments; there is a strong tendency toward lectures based exclusively on textbooks (...) teaching is generally decontextualized from everyday life and science history; critical thinking is not promoted and students have limited

> opportunities to speak or write about science phenomena. (Ministry of Education, 2007, p. 18)

This disconnect is especially worrying, since it is argued that scientific education for citizenship demands a profound revision of its traditional pedagogical approaches.

Researchers in secondary education also point to the increasing difficulties schools encounter to foster meaningful learning experiences and offer a relevant educative proposal (Terigi, 2012). Not surprisingly, as discussed above, Argentine students have demonstrated low proficiency levels in science in national and international examinations which assess student competencies and critical thinking (OECD, 2013; UNESCO, 2014). In addition, very few secondary students express their interest in science and scientific or technological professional careers (Polino, 2012).

We therefore consider that an important challenge of educating for scientific literacy in the country involves a transformation of practices from a teacher-centered to a student-centered approach, not only at school, but also at the teacher preparation level. There is an urgent need to overcome the traditions of "rote learning of scientific contents, with a decontextualized understanding of science, away from everyday life, and unrelated to the historical aspects of science, with little development of scientific skills and critical thinking" (Ministry of Education, 2008, p. 25).

Along these lines, it has been pointed out that initial stages of teacher education often promote the replication of a type of decontextualized school instruction, which is dissociated and overemphasizes content knowledge (Echeverría, 2010; Northfield, 2003). In turn, this approach models the future teachers strategies, creating an environment in which—through their own experiences as learners—pre-service teachers later become reluctant to adopt alternative ways of teaching (Loughran, 2007).

Moreover, we believe that the need to prepare teachers to make scientific content relevant to the students' lives does not only imply enriching pedagogical strategies within teacher education programs, but also involves embedding science teaching (and science teacher education) within a more comprehensive understanding of science.

As research points out, fostering a deeper reflection on the nature of science and its incidence on teaching, as well as enhancing future teachers' understanding of epistemological and historical aspects of science as a human endeavor is, perhaps, one of the biggest debts of the science teacher education system (Adúriz Bravo, 2009). To this regard, student teachers should engage in debates and activities that explore socio-scientific issues, which may provide better understandings of the broad purposes of science as a human endeavor, as well as the value of science learning for the construction of citizenship (European Commission, 2015; Lederman et al.,

2014). In addition, we consider that further development of pedagogical content knowledge and pertinent teaching strategies is crucial to provide student teachers with the necessary tools to engage their future students in active learning and inquiry-based activities.

Nevertheless, a critical challenge to address these changes is still the preparation of teacher educators who, as we pointed out, in many cases have not had personal experiences in research, neither in the natural sciences nor in education. Research has reported that, in many cases, teacher educators hold a distorted and limited idea of the nature of science themselves and have limited capacities regarding the enactment of science practices, as well as a limited understanding of the connections between science, technology and society and how to integrate a more complex vision of science into teaching (Vilches & Gil-Pérez, 2007). Acknowledging this challenge, recent efforts have been undertaken by the INFD in order to offer teacher educators the possibility of engaging in educational research projects, by providing institutional grants and paid time for that purpose, as well as some extra preparation (Serra, 2010). Still, these processes have posed important challenges to teacher educators, who often do not have prior experience in research projects and lack the methodological knowledge needed to design research projects.

In all, we consider that Argentina shows an interesting landscape regarding teacher education, with huge recent efforts undertaken to create a more coherent and articulated system in order to provide a better preparation for all future teachers, including science educators. However, we also believe that there is still an important need to continue revising and improving some fundamental aspects of the system, including efforts to overcome its fragmentation and raise the local capacities of institutions and teacher educators' teams, recognizing them as key actors in the process of raising the quality and equity of science education for all and fostering educational innovations. More specifically, we have identified the need to build a mutually enriching relationship between the teacher education system and the schools, to promote a more comprehensive view of science and to foster the development of a wider and more relevant range of teaching strategies across science teacher education programs. Only then, the goal of quality science education for all will be more than an utopia.

APPENDIX
Jurisdictional Curricular Framework for Secondary Level Biology Teacher Education Programs—Province of Córdoba

	General Education	Specific Education	Professional Exercise and Teaching Practice	Institutional Definition
1st Year	Pedagogy Socio-anthropological issues in education Digital and audio-visual languages	Mathematical models for science Physics models for science General Chemistry General Biology Cellular and molecular Biology	Teaching practice I: contexts and educative practice	
2nd Year	Psychology and education General Pedagogy	Philosophy of science Experimentation in Biology Organic chemistry and Biology Animal Biology I Plant Biology I Biology of microorganisms and fungi	Teaching practice II: Schools, documented stories and everyday life	
3rd Year	History of Education and policy in Argentina	Subjects of education and ESI Science pedagogy History and epistemology of Biology Animal Biology II Plant Biology II Ecology Human Biology	Teaching practice III: the classroom: a place to teach and learn	2 curricular units left for institutional definition
4th Year	Ethics and citizenship Problems and challenges in education	Science pedagogy: biology Genetics and biotechnology History of life on Earth and evolution Environmental education Human biology and health	Teaching practice IV and teaching placement	

REFERENCES

Adúriz-Bravo, A. (2009). *Saberes que circulan en los profesorados de ciencias naturales para el nivel secundario. Informe final.* [Circulating knowledge in secondary science teacher education programs. Final Report] Buenos Aires, Argentina: Centro de Formación e Investigación en Enseñanza de las Ciencias-CeFIEC. Universidad de Buenos Aires.

Aguerrondo, I., & Pogré, P. (2001). *Las instituciones de formación docente como centros de innovación pedagógica.* [Teacher education institutions as centres for pedagogical innovation]. Buenos Aires, Argentina: Troquel.

Cámpoli, O. (2004). *La formación docente en la República Argentina.* [Teacher Education in Argentina]. Buenos Aires, Argentina: Instituto International para la Educación Superior en America Latina (IESALC), UNESCO.

Coordinación General de Estudio de Costos del Sistema Educativo (CGECSE). (2014). *Informe indicativo de salarios docentes Período Octubre-Diciembre 2013.* [Teachers' salaries report] La Plata, Argentina: Ministerio de Educación, Subsecretaría de Planeamiento Educativo.

Cofré, H., González-Weil, C., Vergara, C., Santibáñez, D., Ahumada, G., Furman, M., . . . Pérez, R. (2015, January 30). Science teacher education in South America: The case of Argentina, Colombia and Chile. *Journal of Science Teacher Education, 26*(1), 45–63. http://doi.org/10.1007/s10972-015-9420-9

Davini, M. C., & Alliaud, A. (1995). *Los maestros del siglo XXI. Un estudio sobre el perfil de los estudiautes de magisterio* [Teachers of the 21st century. A study on the profile of teachers' studies]. Buenos Aires, Argentina: Miño y Dávila Editores.

DiNIECE. (2006). *Censo Nacional de Docente. Resultados Definitivos.* [National Teacher Census. Definite Results]. Retrieved from http://diniece.me.gov.ar/index.php?option=com_content&task=category&sectiond=2&id=9&Itemid=20

DiNIECE. (2010). *Operativo Nacional de Evaluación. Censo de finalización de la educación secundaria. Informe de resultados.* [National Assessment Program. Secondary culmination census. Report on Results]. Retrieved from http://diniece.me.gov.ar/images/stories/diniece/evaluacion_educativa/nacionales/resultados/Resultados%20Censo%20ONE%202010.pdf

Echeverría, P. (2010). El papel de la docencia universitaria en la formación inicial de Profesores [The role of university teaching in initial formation teachers]. *Calidad en la Educación, 32*(July), 150–165.

European Commission. (2015). *Science education for responsible citizenship.* Luxembourg: Publications Office of the European Union.

Instituto Nacional de Estadística y Censos (INDEC). (2014). *Encuesta Permanente de Hogares* (EPH) [Permanent housing survey]. Retrieved from http://www.indec.gov.ar/uploads/informesdeprensa/EPH_cont_3trim14.pdf

Instituto Nacional de Formación Docente (INFD). (2013). *Función de investigación en la formación docente. Marcos normativos docentes.* [Investigation in teacher education. Legal frameworks] Buenos Aires, Argentina: Área de Investigación Dirección Nacional de Formación e Investigación.

Esteve, J. M. (2006). Identidad y desafíos de la condición docente [Identity and challenges of the teaching profession]. In Tenti Fanfani, E. (Ed.), *El oficio del docente. Vocación, trabajo y profesión en el siglo XXI.* Buenos Aires, Argentina: Siglo XXI/IIPE-UNESCO/Fundación OSDE.

Ferrata, H. (2014). *Estado de situación de la renovación curricular de la formación docente inicial. Informe del Área de Desarrollo Curricular* [Current status of the update of teacher education guidelines]. Buenos Aires: Instituto Nacioanl de Formación Docente, Dirección Nacional de Formación e Investigación, Área de Desarrollo Curricular.

Furman, M., & Podestá, M. E. (2009). *La aventura de enseñar ciencias naturales* [The adventure of teaching science]. Buenos Aires, Argentina: Aique.

Guadagni Alieto, A., Lasanta, T. I., & Álvarez, M. C. (2014). La secundaria mejora, pero con más desigualdad [Secondary level improves, but with greater inequalities]. *Centro de Estudios de la Educación Argentina, 3*(29), 1–14.

Hisse, M. C. (Ed.). (2014). *Diseño Currricular. Profesorado de Educación Secundaria en Biología Química y Física* [Curricular guidelines. Secondary teacher education in biology, chemistry and physics]. Córdoba, Spain: Ministerio de Educación.

Hisse, M. C. (Ed.). (2009). *Profesorado de Educación Primaria.* [Primary Teacher Education]. Buenos Aires, Argentina: Ministerio de Educación.

Jacinto, C. (2009). *Consideraciones sobre estrategias de inclusion con calidad en la escuela secundaria.* [Strategies for quality educational inclusion at the secondary school] Buenos Aires, Argentina: SITEAL, IIPE-UNESCO Sede Regional Buenos Aires.

Lederman, N. G., Antink, A., & Bartos, S. (2014). Nature of science, scientific inquiry, and socio-scientific issues arising from genetics: A pathway to developing a scientifically literate citizenry. *Science & Education, 23*(2), 285–302.

Loughran, J. J. (2007). Science teacher as learner. In S. K. Abell & N. G. Lederman (2007). *Handbook of research on science education* (pp. 1043–1065). Mahwah, NJ: Lawrence Erlbaum.

Loughran, J. (2014). Professionally developing as a teacher educator. *Journal of Teacher Education, 65*(4), 271–283.

Mezzadra, F., & Veleda, C. (2014). *Apostar a la docencia. Desafíos y posibilidades para la política educativa argentina.* [Betting on teaching. Challenges and opportunities for educational policy in Argentina]. Buenos Aires, Argentina: Fundación CIPPEC.

NGSS Lead States. (2013). *Next Generation Science Standards: For States, By States.* Washington, DC: The National Academies Press.

Northfield, J. R. (2003). Teacher educators and the practice of science teacher education. In B. Fraser & K. Tobin (Eds.), *International handbook of science education* (pp. 695–706). Dordrecht, The Netherlands: Kluwer.

Organization for Economic Development and Cooperation (OECD). (2014a.). *PISA 2012 Results: What Students Know and Can do: Student Performance in Reading, Mathematics and Science (Volume I, Revised edition).* OECD Publishing. Retrieved from http://www.oecd.org/pisa/keyfindings/pisa-2012-results-volume-i.htm

Organization for Economic Development and Cooperation (OECD). (2014b). *PISA 2012 results in focus. What 15-year-olds know and what they can do with what they know.* Retrieved from http://www.oecd.org/pisa/keyfindings/pisa-2012-results-overview.pdf

Organization for Economic Development and Cooperation (OECD). (2013). *Panorama de la educación 2013. Indicadores de la OCDE.* [Education Panorama 2013] Santillana, Spain: Retrieved from http://www.oecd.org/edu/Panorama%20de%20la%20educacion%202013.pdf

Pasmanik, D., & Cerón, R. (2005). Las prácticas pedagógicas en el aula como punto de partida para el análisis del proceso enseñanza-aprendizaje: Un estudio de caso en la asignatura de química [Pedagogical practices in the classroom as a starting point for analyzing the teaching-learning process: A case study in the subject of chemistry]. *Estudios Pedagógicos, 31*(2), 71–87.

Perazza, R., & Terigi, F. (2008). Decisiones políticas acerca de la evaluación docente. Consideraciones sobre la experiencia y la reformulación de la evaluación de desempeños en la Ciudad de Buenos Aires. [Political decisions on teacher assessment]. *Revista Iberoamericana de Evaluación Educativa, 1*(2), 23–40.

Perlo, C. (1998). *Hacia una didáctica de la formación docente* [Towards a didactics of teacher education]. Buenos Aires, Argentina: Homo Sapiens Ediciones.

Perrenoud, P. (2004). *Desarrollar la práctica reflexiva en el oficio de enseñar: Profesionalización y razón pedagógica.* [Developing reflexive practice in teaching: Professionalization and pedagogical fundaments] Barcelona, Spain: Graó.

Pogré, P. (2010). *Proyecto de Mejora para la Formación Inicial de Profesores para el Nivel Secundario. Áreas: Biología, Física, Matemática y Química.* [Secondary Teacher Education Improvement Proyect. Biology, Physics, Math and Chemistry]. Ministerio de Educación de la Nación, Instituto Nacional de Formación Docente y Secretaría de Políticas Universitarias. Retrieved from http://repositorio.educacion.gov.ar/dspace/handle/123456789/89786

Polino, C. (2012). Las ciencias en el aula y el interés por las carreras científico-tecnológicas: Un análisis de las expectativas de los alumnos de nivel secundario en Iberoamérica [Interest in science and scientific carriers: An analysis of secondary students expectations in Iberoamerica]. *Revista Iberoamericana de Educación, 58*(2012), 167–191.

Sequeira, J. (2009). *Segundo estudio regional comparativo y explicativo. Aportes para la enseñanza de las ciencias naturales. UNESCO* [Second regional comparative and explicatory study. Insights for science teaching]. Santiago, de Chile: Salesianos Impresores.

Serra, J. C. (Dir.). (2010). *Estado de situación de la investigación en los institutos de formación docente. Informe Final.* [Current status of investigation in teacher education institutes. Final Report]. Buenos Aires, Argentina: Ministerio de Educación de la Nación.

Shulman, L. S. (1986). Those who understand: Knowledge growth in teaching. *Educational researcher, 15*(2), 4–14.

Tedesco, J. C. (2007). ¿Para qué educamos hoy? [What do we teach for today?] In I. Dussel & P. Pogré. (Eds.). (2007). *Formar docentes para la equidad.* (pp. 13–28). Buenos Aires, Argentina: Propone.

Tenti Fanfani, E. (Coord.). (2009). *Abandono escolar y políticas de inclusion en la educación secundaria* [School drop out and inclusion policies in secondary education]. Buenos Aires, Argentina: Programa Naciones Unidas para el Desarrollo–PNUD.

Tenti Fanfani, E. (2005). *La condición docente: Análisis comparado de la Argentina, Brasil, Perú y Uruguay.* [The teaching profession: Comparative analysis of Argentina, Brasil, Perú and Uruguay]. Buenos Aires, Argentina: Fundacion Osde.

Terigi, F. (2012). La cuestión curricular en la educación secundaria. [Curriculum in secondary education] In E. Tenti Fanfani (coord.), *La escolarización de los adolescentes: Desafíos culturales, pedagógicos y de política educativa* (pp. 55–77). Buenos Aires, Argentina: IIPE-UNESCO.

UNESCO. (2012). *Antecedentes y criterios para la elaboración de políticas docentes en América Latina y el Caribe. Proyecto Estratégico Regional sobre Docentes "Profesores para una Educación para Todos"* [Background and criteria to elaborate teacher policies in Latinamerica and the Caribean]. Santiago, de Chile: Oficina Regional de Educación de la UNESCO para América Latina y el Caribe OREALC UNESCO.

UNESCO. (2014). *Primera entrega de resultados. Tercer Estudio Regional Comparativo y Explicativo (TERCE)* [First delivery of results. Third Regional Comparative and Explicatory Study]. Santiago, de Chile: Oficina Regional de Educación de la UNESCO para América Latina y el Caribe OREALC UNESCO.

Vaillant, D. (2013). Formación inicial del profesorado en América Latina: Dilemas centrales y perspectivas. [Initial teacher training in Latin America: Central dilemmas and perspectives]. *Revista Española de Educación Comparada, 22*(2013), 185–206.

Vilches, A., & Gil-Pérez, D. (2007). La necesaria renovación de la formación del profesorado para una educación científica de calidad. [The necessary renovation of teacher education for quality science education]. *Tecné, Episteme y Didacxis,* 22(#), 67–85.

Legislation and Other Official Documents

Federal Law of Education. (1993). N° 24.195.

Law of Higher Education. (1995). N° 24.521.

National Law of Education [LEN]. (2006). N° 26.206.

Draft Bill (2015). N° 1691/15. Presented at Congress 18.05.2015, pending approval. Retrieved from http://www.senado.gov.ar/parlamentario/comisiones/verExp/1691.15/S/PL

Federal Board of Education. (2004). Resolution N° 214/04.

Federal Board of Education. (2007b). Resolution N° 23/07 Annex I. Plan Nacional de Formación Docente 2007–2010.

Federal Board of Education. (2007d). Resolution N° 24/07 Annex I. "Lineamientos Curriculares Nacionales para la Formación Docente Inicial."

Federal Board of Education. (2007f). Resolution N° 30/07 Annex I. Hacia una institucionalidad del Sistema de Formación Docente en Argentina.

Federal Board of Education. (2011b). Resolution N° 140/11 Annex I. "Lineamientos Federales para el planeamiento y organización institucional del sistema formador."

Federal Board of Education. (2011c). Resolution N° 134/11. Responsabilidades de los ministerios nacional y jurisdiccionales en la Educación Inicial, Primaria, Secundaria y de Formación Docente.

Federal Board of Education. (2012b). Resolution N° 167/12 Annex I. Plan Nacional de Formación Docente 2012–2015.

Ministry of Education. (2007). *Comisión Nacional para el mejoramiento de la enseñanza de las Ciencias Naturales y la Matemática. Informe Final Agosto 2007.* [National Commision for the improvement of Science and Math teaching] Retrieved from http://www.me.gov.ar/doc_pdf/doc_comision.pdf

Ministry of Education. (2008). *Mejorar la enseñanza de las ciencias y la matemática: Una prioridad nacional. Informe y recomendaciones de la Comisión Nacional para el mejoramiento de la enseñanza de las ciencias naturales y la matemática* [Improving science and math education: a national priority]. Retrieved from http://portal.educacion.gov.ar/ files/2009/12/Mejoramiento-de-la-ense%C3B1anza.pdf

Ministry of Education, Science and Technology. (2012). *Lineamientos básicos sobre formación docente de profesores universitarios elaborados por la commission mixta ANFHE-CUCEN* [Basic guidelines for university teacher education elaborated by the ANFHE-CUCEN commission]. La Plata, Argentina: Author.

CHAPTER 11

SECONDARY SCIENCE TEACHER EDUCATION IN AUSTRALIA

Mihye Won
Curtin University

Mark Hackling
Edith Cowan University

David F. Treagust
Curtin University

In this chapter, we describe and discuss initial teacher education programs in Australia for secondary science as well as professional standards for continuing science teacher education. Beginning with a brief overview of secondary science teacher education reform, we discuss the administration related topics, such as admission to teacher education programs, program accreditations, and science teacher registrations. Next, we describe the required structure and content of initial science teacher education programs, with an emphasis on content knowledge (CK), pedagogical knowledge (PK), and professional practice. We describe five priority areas for

Model Science Teacher Preparation Programs, pages 229–248

Australian teacher education, including the education of Indigenous students, classroom management, literacy and numeracy education, the role of information and communications technologies (ICT) in teaching and learning, and the education of students with special needs. The chapter also describes expectations regarding the content and PK for practicing secondary science teachers and the factors driving changes and enhancements to the knowledge required for effective teaching of science. We close by considering the issue of out-of-field teaching which is a particular challenge for remote and other hard-to-staff schools.

PUSH FOR FEDERAL REGULATIONS ON THE SECONDARY TEACHER EDUCATION REFORM

Australian politicians and the general public have often talked about the "problems" and the "quality" of school education in recent years. They start such discussions with the recent international student achievement test scores, such as PISA or TIMSS (Australian Institute of Teaching and School Leadership, 2013; NSW Department of Education, 2013; Pyne, 2014). While Australian students performed reasonably well in those tests (above the OECD averages), it is emphasized that mean scores of Australian students have not improved much, and they are significantly lower than several countries including Japan, Finland, Estonia, and Korea (Thomson, De Bertoli, & Buckley, 2013). In addition, the number of students taking advanced level mathematics and science subjects in the last years of secondary school is decreasing (Kennedy, Lyons, & Quinn, 2014). This is deemed to lead to a lack of quality scientists to compete and lead the world economy (Office of the Chief Scientist, 2012). Because teachers are recognized as the single most important factor that influences students' performance (Hattie, 2009), it has been heatedly debated how to improve the quality of science teachers and science teaching. The Teacher Education Ministerial Advisory Group (TEMAG) was established in 2014 (and the report released on February 13, 2015) to review the initial teacher education programs and provide recommendations to strengthen the consistent application of the national standards for initial teacher education across Australia.

Australia has a federal political system in which the national government has the majority of tax-raising powers whilst the states have most responsibility for school education. Consequently, the states have been reluctant to give up their powers for educational policy whilst the national government as the main funder of education has sought greater control. The Australian political climate currently supports greater national consistency in educational policy, and Australia is initiating a number of educational reforms to this purpose. These reforms include: a national school curriculum, national teacher professional standards, national standards for the accreditation of initial teacher education (ITE) programs and for the registration of teachers.

These initiatives have been driven by the establishment of two key national education bodies: the Australian Curriculum, Assessment, and Reporting Authority (ACARA) which is responsible for school curriculum and for monitoring educational standards; and the Australian Institute for Teaching and School Leadership (AITSL) which is responsible for professional standards. These two bodies report to the Education Council (part of the Council of Australian Government) so that federal and state ministers of education have joint oversight of national education policy development.

ADMISSION, ACCREDITATION, AND REGISTRATIONS

Admission to Teacher Education Programs

Currently, Australian universities make their own decisions regarding admission standards for ITE programs and for graduating students upon successful completion of ITE courses. A recent review of Australian ITE recommends that ITE providers "select the best candidates into teaching" (TEMAG, n.d., p. xv). This recommendation comes from the public concern about the low entry standards for teacher education programs. Science and engineering majors tend to have top-performing students while education majors consist of a minimal number of high academic achievers compared to any other majors (Ferrari & Rushton, 2014). If teachers themselves did not do well in school learning, will they be able to inspire students to learn and excel in school (Jones, 2014)? However, teacher educators have asserted that the best teacher candidates need to have personal attributes as well as academic skills to become successful teachers, and TEMAG acknowledge that ITE providers need to use a combination of academic achievement data and candidates' personal attributes in their selection process.

TEMAG also recommends ITE programs include "final assessments that ensure pre-service teachers are classroom ready" (TEMAG, n.d., p. xii), which might include a portfolio of evidence to demonstrate their achievement of the national professional standards for graduate teachers. Louden (2015) reports that new standardized assessments of professional literacy and numeracy have been prepared for AITSL and are

> designed to assist teacher education providers in measuring the standards of literacy and numeracy specified in the national teacher education accreditation Program Standard 3.2, which requires that [teacher education] students achieve a standard equivalent to the top 30% of the population by the time they graduate. (p. 35)

Accreditation of Teacher Education Programs

Almost all secondary teachers complete their initial teacher education (ITE) at universities which have been self-accrediting institutions; that is, they have responsibility for the content and standards of teacher education programs. The self-accrediting nature of Australian universities has led to great diversity and standards of ITE programs. According to the Initial Teacher Education: Data Report (AITSL, 2014), we have over 400 different ITE programs at 48 providers for 78,000 students across Australia in 2012.

With the advent of AITSL in 2010, the release of the *Australian Professional Standards for Teachers* (AITSL, 2011), and the *Accreditation of Initial Teacher Education Programs in Australia: Standards and Procedures* policy (AITSL, 2013), nationally consistent standards have been developed for the accreditation of ITE programs and registration of teachers. The professional standards, shown in Table 11.1, are for professional knowledge (two items), professional practice (three items), and professional engagement (two items). The professional standards are differentiated into career stages of teacher professional development; namely, graduate teachers, proficient teachers, highly accomplished teachers, and lead teachers. All ITE programs are expected to ensure that their graduates achieve the seven professional standards for graduate teachers.

Whilst the policy is a national one, it has been implemented by state-based teacher registration boards. For Western Australia, for example, it is called the Teacher Registration Board of Western Australia (TRBWA). There are minor differences between state teacher registration boards in their requirements for registration. TEMAG (n.d.), in a recent document, recommends that AITSL become a stronger national regulator of ITE programs and manage the accreditation processes. In consultation with the

TABLE 11.1 The Australian National Professional Standards for Teachers (AITSL, 2011)

Professional Knowledge
1. Know students and how they learn
2. Know the content and how to teach it
Professional Practice
3. Plan for and implement effective teaching and learning
4. Create and maintain supportive and safe learning environments
5. Assess, provide feedback and report on student learning
Professional Engagement
6. Engage in professional learning
7. Engage professionally with colleagues, parents/carers and the community

state teacher registration boards, the AITSL is slowly putting more of the accreditation standards into action.

Teacher Registration

There is no national qualifying examination to become a science teacher in Australia. However, a recent review by TEMAG suggests that a compulsory test on teacher candidates' literacy and numeracy skills will be a requirement for graduation (TEMAG, 2015).

Once students successfully complete their university studies on secondary science education, they must be registered by the state teacher registration board to be able to teach. Registration requires appropriate qualifications, experience, and police checks to ensure the person is suitable for the profession. Teachers are normally given provisional registration (of graduate teachers), and after a period of service of 3–5 years, they need to provide evidence that they have acquired the knowledge, skills, and attitudes of proficient teachers and have met the professional standards of proficient teachers. For full registration, teachers are assessed by the state teacher registration board. With further experience and professional learning, teachers can be assessed and accredited as Highly Accomplished Teachers and as Lead Teachers. The AITSL provides the national professional standards and descriptors for teachers at each stage of their careers as noted above.

STRUCTURE OF NATIONAL REQUIREMENTS FOR INITIAL SCIENCE TEACHER EDUCATION PROGRAMS

While the state teacher registration board (e.g., TRBWA) is responsible for the accreditation of ITE programs in its state and the registration of teachers, AITSL (2013) provides guidelines for the structure and content of ITE programs. The ITE accreditation policy by AITSL also provides broad guidelines about the amount of discipline studies, professional studies, and practicum that should be included in ITE programs.

ITE programs for the education of secondary science teachers must be at least four years in duration at the university level. There are three pathways: a four-year Bachelor of Education (BEd) degree which integrates discipline studies, professional studies, and practicum; a three-year Bachelor of Science degree (BSc) followed by one year of professional education (graduate Diploma of Education; DipEd); and a BSc followed by two years of professional education (Master of Teaching; MTeach). The one-year professional DipEd is allowed under the current accreditation policy, and TRBWA is currently re-accrediting the Western Australian Graduate Diplomas of Education. However, the DipEd is being phased out in many universities and replaced with the two-year professional graduate degree, the MTeach, following the AITSL guidelines. The existing pathways are structured similarly but the DipEd or MTeach is without studies in the subject content areas. The difference between the BEd in Secondary Education (Science) and MTeach can be seen in Tables 11.2 and 11.3. For reference, full-time students usually take four courses per semester. According to the government's workforce analysis (Harris, Jensz, & Baldwin, 2005), the older generation of science teachers tends to have a BSc degree with a DipEd while younger generations of science teachers tend to have a BEd degree.

GUIDELINES FOR COMMON CORE REQUIREMENTS

AITSL specifies common core requirements for initial secondary science teacher education programs related to CK, PK, and professional practice (practicum).

TABLE 11.2 A Sample Course Structure of Bachelor of Secondary Education (With Physics Major & Mathematics Minor)

# of Courses	Domain Area	Organizing Faculty
10	Core education studies (general pedagogy courses)	Education
6	Major content studies (e.g., physics)	Science
2	Major subject area curriculum & pedagogy (e.g., physics teaching methods)	Education
4	Minor content studies (e.g., mathematics)	Science
1	Minor subject area curriculum & pedagogy (e.g., mathematics teaching methods)	Education
3	Practicum	School

TABLE 11.3 A Sample Course Structure of Master of Teaching in Secondary Education

# of Courses	Domain Area	Organizing Faculty
10	Core education studies	Education
2	Major curriculum & pedagogy studies	Education
2	Minor curriculum & pedagogy studies	Education
4	Practicum (15 days each)	School
1	Professional learning or research project	Education

Science Content

Australian teacher candidates select two subject areas (major teaching area and minor teaching area). The major teaching area refers to teaching years 7–12 (middle and high school level science subjects, e.g., physics and chemistry). The minor teaching area refers to being able to teach years 7–10 (middle school level science). To qualify as a science major teaching area, the teacher education program students need to take six discipline specific science courses (e.g., physics or chemistry) with no more than two courses at first-year level science and at least two at third- or fourth-year science major courses. The science courses need to be taken over three years. To qualify as a science minor teaching area, the teacher candidate requires four discipline-specific science units, with no more than two at first-year level science courses. Given that the national curriculum for science in lower secondary comprises biological, chemical, physical and earth-space sciences, it is expected that the discipline studies include units from across the four science strands with more in-depth studies of one or two of these strands that might be taught at the senior secondary level.

Secondary Science Teaching Methods

In addition to the discipline-specific content studies, a teacher candidate needs to take both lower secondary and upper secondary science teaching methods courses (curriculum & instruction) for the discipline. The lower secondary teaching method courses is on teaching of middle school level sciences while the upper secondary curriculum and instruction course focuses on their specialized science discipline. Students usually specialize in a science "discipline" that relates to the school curriculum; for example, physics or chemistry. To qualify as a science minor teaching area, the students additionally need to take a curriculum and instruction course specializing in lower secondary science education. Most universities do not allow the same major and minor teaching areas (for example, pre-service teachers cannot have a science major and science minor).

Professional Studies

A teacher candidate needs to take at least 10 general education courses (see Tables 11.2 and 11.3). Universities usually include courses such as general introduction to secondary teaching, educational psychology or child development, learning theory, assessment, curriculum, teaching literacy, educational technologies, inclusive education, indigenous Australian education, and reflective teaching (teacher as researcher).

Professional Field Experience

ITE programs include a teaching practicum in schools linked through enduring partnerships with the university.

> The professional experience component of each program must include no fewer than 80 days of well-structured, supervised and assessed teaching practice in schools in undergraduate and double-degree teacher education programs and no fewer than 60 days in graduate entry programs. (AITSL, 2013, p. 14)

The field experience (practicum) is usually divided into two separate phases. For the first three years of a bachelor of secondary education program or for the first three semesters of the MTeach for secondary science preservice teachers, professional field experience occurs around three weeks each year/semester with the expectation of an increasing practicing teaching load (e.g., from 1–2 hours per day to 3–4 hours per day). For the final year or final semester of the ITE programs, the education internship lasts for 10 weeks. For the internship, students are expected to take over the full teaching load of the mentor teachers (20 hours per week) in major and minor teaching areas. Some universities require teacher education program students go to school every semester for 5 days for the first 3 years, and during the fourth year they go to their placement schools for 25 days for each semester.

During the professional field experience, mentor teachers maintain the duty of care of the pre-service teachers, providing feedback on lesson plans and actual lessons based on the university-provided guidelines and national professional standards. University-based supervisors liaise with the mentor teachers and observe each teacher candidate's teaching at least once per placement to provide constructive feedback.

National Priority Areas

AITSL's (2013) national program standards for initial teacher education identifies five priority areas to include in teacher education programs: Aboriginal and Torres Strait Islander education (indigenous education), classroom management, ICT (information and communication technologies), literacy and numeracy, and students with special educational needs. Initial teacher education programs may have separate units for these priority areas or embed them across the program of study.

Those priority areas, except the ICT component, are consistently identified as the difficulties that early career teachers experience within the first five years of their teaching careers in secondary school. According to the national survey (McKenzie, Weldon, Rowley, Murphy, & McMillan, 2014), early career teachers and their employers identified the areas where they found ITE programs not very helpful. These areas included teaching students from indigenous backgrounds, working effectively with parents and guardians, teaching students with learning difficulties, teaching students from different cultural backgrounds, and developing students' literacy and numeracy skills.

Aboriginal and Torres Strait Islander Education

Aboriginal and Torres Strait Islander children experience social and educational disadvantage, and many are not engaged with schooling or learning, which results in significantly lower levels of educational attainment (Trewin & Madden, 2005). Poor educational outcomes are a cause of decreased employment opportunities and are correlated with poor health, increased engagement in risky behaviors, lower life expectancy, and increased likelihood of negative encounters with the justice system (Beresford, Partington, & Gower, 2012). In their review of the cultural factors impacting the science education of Indigenous students, Hackling, Byrne, Gower, & Anderson (2015) note:

> ... that patterns of behaviour, language use, use of space and time that characterise the culture of schooling reflect the values and beliefs of the dominant culture. This means that students from non-dominant groups must take up the behaviours and language of a second culture to be successful in school. An additional cultural border crossing involves accommodating to the culture of school science with its own language and ways of generating, testing and representing ideas. Negotiating the meaning of science concepts where there are no equivalent concepts in students' first languages and cultures provides an additional barrier to learning science (Chigeza, 2008). Instruction that supports students from non-dominant cultural groups to be successful in school science needs to value students' language and cultural backgrounds, take account of the nature of science and connect science to the experiences of students (Lee, 2004). Without this cultural connection to school science, student engagement will be limited. Gower and Byrne (2012) consider teachers' cultural competence as the capacity to support Aboriginal students to navigate the cultural border crossings between Indigenous and non-Indigenous cultures and knowledge systems. (p. 28)

The majority of teacher education candidates are not from an Indigenous background; thus, non-Indigenous teacher education candidates need to develop knowledge of Indigenous Australians' history and culture through their initial teacher education programs. They need to be able to identify effective teaching and communication strategies that fit the learning needs of Indigenous students and to connect with their families and communities. To ensure the effectiveness of these teaching and communication strategies, AITSL (2013) requires appropriately qualified Aboriginal people be involved in the design, delivery, and evaluation of the cultural content to represent their perspectives.

In addition, there have been calls to increase the effort to recruit and retain Indigenous preservice teachers. Pechenkina and Anderson (2011) report Indigenous Australians' low participation and high attrition rates in higher education. Some universities actively adopt alternative pathways to encourage students with Indigenous backgrounds to pursue higher education (Pechenkina & Anderson, 2011). AITSL (2013) also prescribes that Indigenous students in teacher education programs are supported through local Indigenous support units.

Classroom Management

Many beginning teachers express difficulty with classroom management because the construction of effective and productive learning environments demands experience in diverse human interactions. Effective teaching strategies required by teacher candidates include connecting with youth, dealing with common classroom misbehaviors, and restorative processes, as well as learning how to de-escalate conflict (McDonald, 2013). Teacher candidates need to develop their knowledge in students' cognitive, emotional, and behavioral development and promote positive attitudes towards learning and their peers, minimizing antisocial behaviors. Teacher candidates also need to be ready to facilitate strategies to communicate effectively with students and their parents/guardians.

Information and Communication Technologies

The Australian Federal Government's Digital Education Revolution (DER) national strategic plan envisaged that all students in year 9 to 12 classrooms would have 1:1 access to computers. Since this announcement, there has been more than $2 billion invested in schools' ICT resources (Ministerial Council on Education Employment Training and Youth Affairs, 2012). Koehler and Mishra (2009) propose that Shulman's model of pedagogical content knowledge (PCK) needs to be expanded to take account of the knowledge required for teaching with ICTs. Their technological pedagogical content knowledge framework (TPACK) comprises Shulman's three primary knowledge domains of CK, PK, and PCK but extends these knowledge domains to include technology knowledge (TK) as a way to understand the role of technology in the process of teaching and learning. Research consistently indicates that "many teachers are not integrating these tools into their instruction in ways that support and maximize student learning" and "[c]onsequently, there is need to motivate, train, and equip teachers with the skills, knowledge, and pedagogical framework to effectively teach with technology tools" (Keengwe & Onchwari, 2011, p. 1). Teacher candidates, therefore, need to build a sophisticated understanding and practical knowledge of the integration of ICTs into education and to be able to critically evaluate its social and pedagogical implications.

Literacy and Numeracy

Students in teacher education programs need to understand the literacy and numeracy aspects in the curriculum area and provide appropriate support for students to build the reading, writing, speaking, and listening capabilities within the subject areas. Given that science has particular ways of representing and communicating scientific information, teachers of secondary science need to develop the subject-specific literacies of science and know how to support students to use these to access and communicate information effectively (Lemke, 1990). Given concerns about the personal and professional literacies of teaching candidates, AITSL is currently piloting online literacy and numeracy tests that all student teachers will be required to pass as a condition for graduating from ITE programs.

Special Learning Needs of Students

AITSL (2013) notes that "[i]n inclusive classrooms graduate teachers will work with the full spectrum of children in terms of abilities, difficulties and special needs, including learning difficulties and giftedness" (p. D6). Graduate teachers are expected to understand the likely impacts that special needs may have on a student's access to and participation in learning, and with specialist support, be able to work with the child's family to support and engage special needs students in learning. Teachers are expected to have the skills required to "use curriculum based assessment and monitoring procedures for identifying prior achievement and making differentiated assessment and learning decisions, including the reasonable adjustments required to enable students to meet curriculum outcomes" (AITSL, 2013, p. D6). Initial teacher education programs typically offer a unit of study specifically dealing with learning disability, giftedness, and inclusive education.

GUIDANCE FOR CONTENT AND PEDAGOGICAL KNOWLEDGE OF SECONDARY SCIENCE TEACHERS: FROM ITE GRADUATES TO LEAD TEACHERS

The current Australian standards and policies only provide general outlines of the types of knowledge required by all teachers, for example, "knowledge and understanding of the concepts, substance and structure of the content and teaching strategies of the teaching area" (AITSL, 2013, p. 2). Before the AITSL released the professional standards for all teachers, the Australian Science Teachers Association (ASTA) initiated the development of national professional standards for Australian teachers with its National Professional Standards for Highly Accomplished Teachers of Science (ASTA, 2009). Different from the national teacher standards that encompass all teachers, the professional standards by ASTA, specifically for teachers of science, gave more detailed guidance about the knowledge required for teaching. For example, the standard relating to knowledge of science addresses both conceptual knowledge of the disciplines and knowledge about the nature of science and inquiry:

> Highly accomplished teachers of science understand the nature and dimensions of science: [T]hat it is a body of knowledge, a way of thinking and communicating, asking and answering questions, and of interpreting events and phenomena from scientific perspectives. They are confident in their knowledge of the major ideas of the disciplines of science related to their area of teaching: the fundamental principles, laws, theories, models and facts and the conceptual themes of cause and effect, patterns of change, systems and interactions, structure and function. They understand the dynamic relationship and diversity of connections that exist between and within the disciplines and dimensions of science, technology and other areas of learning. (ASTA, 2009, p. 5)

ASTA's professional standards also specified the requirements for highly accomplished teachers to understand the nature of science and the processes of inquiry.

> Highly accomplished teachers of science understand the distinctive modes of scientific inquiry: [T]he ways of predicting and knowing, built on observation, objectivity, testing and evidence, that contribute to conceptual understanding and scientific literacy. They understand the interdisciplinary and collaborative nature of current scientific activity; they know how to model the diverse practices of science and habits of mind that constitute scientific inquiry and investigation. They understand the kinds of questions that can be tested scientifically and those that cannot; how to plan and conduct scientific investigations, collect evidence and interpret data. They can identify patterns and trends that inform their thinking and understanding and are able to transpose and communicate information in the language and literacies of science, integrating multimodal forms skilfully. They use scientific terminology and conventions and know how to incorporate these in their teaching. (ASTA, 2009, p. 5)

The ASTA professional standards also specified the PK required of highly accomplished teachers of science, for example:

> Highly accomplished teachers of science know how to draw on a variety of pedagogic strategies for timely diagnosis of their students' understanding and skill-base in science and for intervention to guide learning. Understanding that students view science and science learning through a cultural interpretative framework, they know how to reconcile disparate views and facilitate cross-cultural understanding. They recognise that teacher modelling of scientific thought processes and practices can scaffold learning and make complex concepts accessible. They know how to construct opportunities for their students to investigate their science-related questions independently and collaboratively in formal and informal settings, using diverse resources. (ASTA, 2009, p. 6)

While the ASTA's professional standards include comprehensive guidelines for science teachers in Australia, they do not carry the same weight as the AITSL's professional standards in regards to teacher registrations, promotions, or accreditations.

THE NEED FOR ONGOING ENHANCEMENT OF TEACHER KNOWLEDGE

Professional standards are inevitably written in general terms specifying the types of knowledge required for teaching rather than giving detailed specifications. Clearly, there are unchanging core areas of domain-specific conceptual knowledge, domain-general PK, and some aspects of PK that are domain specific such as those pedagogies related to scaffolding scientific inquiry. However, there are a number of factors that continually shape and change the knowledge required for teaching in particular places and at different times. Given that teaching and learning are culturally framed and strongly influenced by the local context (Alexander, 2000; Hackling, Chen, & Romain, 2017), teachers need to develop their knowledge base within cultural and contextual settings. Given the diversity of educational settings in Australia with metropolitan, regional, rural, and very remote communities with high Aboriginal enrolments, teachers must understand the local community and the particular educational needs of that community. Teaching in these varied settings requires the development of cultural awareness and strategies appropriate to those particular settings.

Educational policy, philosophy, and curriculum change with time: These changes also require adjustments to the teacher's knowledge-base. As educational research progresses, we also develop new insights into effective teaching practice which inform changes to the pedagogical repertoire for the enhancement of teacher effectiveness. The current attention being given to STEM education provides one example of the changing focus of science education which has implications for teachers' practice, and hence, their content and PK. Whether the call for STEM education is a way to emphasize the teaching and learning of the subjects Science, Technology, Engineering, and Mathematics (STEM) in the curriculum or a way to integrate the teaching and learning of these subjects is a moot point given the subject-oriented curriculum and the subject-oriented education of the Australian teaching workforce.

Given the widespread concerns about STEM education (e.g., Houses of Parliament, Parliamentary Office of Science and Technology, 2013; Office of the Chief Scientist, 2012; White House Office of Science and Technology Policy, 2015), especially in Western developed countries, many policy makers are drawing on the latest science education research to inform policy responses that will enhance the teaching of STEM subjects in schools. A number of Australian reports (The New Work Order, 2015; PriceWaterhouseCooper, 2015) are highlighting the importance of higher order thinking and reasoning as critical learning outcomes from STEM education that will enable young people to thrive in digitally disrupted and knowledge-based economies. Research into the role of classroom discourse and multimodal representation is providing valuable new insights into practices that scaffold scientific reasoning drawing on these representational modes.

Classroom discourse research demonstrates that increasing the proportion of open and higher order questions creates higher levels of cognitive engagement. "In classrooms where higher-order questioning was observed, students also engaged at deeper levels with science concepts, formulating hypotheses and using evidence to draw conclusions about phenomenon" (Smart & Marshall, 2013, p. 265). Research has also highlighted the importance of combining appropriate questions with discourse moves that enable teachers to respond to students' answers in ways that are productive (Hackling & Sherriff, 2015). Tytler and Aranda's (2015) analysis of discourse in expert teachers' classrooms show that such discourse moves serve "three broad purposes: to elicit and acknowledge student responses, to clarify and to extend student ideas" (p. 425). It is the combination of the right questions that open up substantive talk and discourse moves that challenge and extend students' thinking in ways that require them to substantiate claims with reasons and to engage them in extended conversational threads that fully explore the students' thinking and science concepts productively.

Discourse is one of many productive ways of representing students' thinking and science concepts. Video ethnographic research in the classrooms of expert teachers is revealing the rich multimodality of learning environments and the range of modalities used to represent and communicate ideas about natural phenomena. Embodied modes such as gesture and role play, symbolic modes such as mathematical equations, and graphical modes such as drawings, all have particular affordances for representing thinking and supporting reasoning as the representations are constructed and then challenged in classroom conversations (Gilbert & Treagust, 2009; Kress, Charalampos, Jewitt, & Ogborn, 2001; Treagust & Tsui, 2013). Pedagogical reasoning of the highest order is required of skilled teachers as they plan teaching sequences that combine a number of different representational modes and orchestrate them into meaningful learning sequences (Hackling, Murcia, & Ibrahim-Didi, 2013).

These examples of effective science teaching pedagogy that create rich opportunities for students to be engaged in the practices of the discipline and, through these practices, engage in reasoning and extending their conceptual understandings can develop some of the learning outcomes that policy makers would like to see emerge from enhanced science education in our schools. Such sophisticated practice requires a rich knowledge-base for teaching and high level pedagogical reasoning. Given the changing emphasis and foci of science education, there is a need for ongoing professional learning that supports teachers as they expand and enrich their knowledge and practice to enable continuous improvement in teaching and learning.

IMPLEMENTING THE PROFESSIONAL STANDARDS AND THE REALITIES OF SMALL REGIONAL SCHOOLS: OUT-OF-FIELD TEACHING

While the ITE providers and professional organizations work towards implementing more effective ITE programs for future science teachers, many Australian schools are experiencing a lack of qualified science teachers, especially in physics and chemistry (Harris et al., 2005). Ainley, Kos, & Nicholas (2008) provide data related to Australian secondary science teacher qualifications as shown in Table 11.4.

These data suggest that a significant number of lower secondary science (56%) teachers are teaching with less than two years of tertiary study in the subject. At the year 8 level, the 2011 TIMSS data show that 55% of the sampled Australian science teachers had a major in science and science education. At the critical phase of transition from elementary education to secondary education, children need to be engaged and inspired to learn science if they are to be retained in the STEM education pathway through year 12 science and beyond. This requires passionate teachers with rich science PCK and is unlikely to occur with teachers who have limited science PCK, confidence, and self-efficacy for teaching science.

This situation is exacerbated for low socioeconomic status (SES) schools or small schools in remote areas. Even with governments' financial incentives to attract and retain qualified teachers in such hard-to-staff schools, science is being taught by "out-of-field" teachers who are currently teaching

TABLE 11.4 Australian Secondary Science Teacher Qualifications

Field of Teaching	Percentage of Teachers With 2 or More Years of Tertiary Study in the Field	Percentage of Teachers With 3 or More Years of Tertiary Study in the Field
Biology Years 11–12	85	78
Chemistry Years 11–12	87	73
Physics Years 11–12	76	60
General Science Years 7/8–10	44	38

science even though they are not qualified to teach science (Australian Government Productivity Commission, 2012).

Teachers and principals in remote, low SES schools recognize the shortage of qualified science teachers as "reality." Instead of dismissing the out-of-field teachers as unqualified science teachers, ITE providers need to look into ways to make such out-of-field teaching an opportunity for new experiences and in this case learning to teach science as part of their work (Hobbs, 2013). For example, ITE providers can teach preservice teachers to develop attitudes of flexibility in their teaching approaches, to learn strategies to gain support from their school for building their expertise in the subject (e.g., time and resource materials), and how to identify and work with other teachers who could serve as subject-specialist mentors (Hobbs, 2013).

THEMES THAT CAN BE DRAWN FROM THE WORK IN THIS CHAPTER

During the past decade, there have been substantial changes in the education of teachers in Australia, and this includes the education of secondary science teachers, the focus of this chapter. We comment on three themes.

The first theme is the progression towards national consistency in school curriculum, teacher professional standards, teacher education standards, and accreditation of teacher education programs. Within these emerging national frameworks, there remains great diversity within ITE programs offered by universities. To date, university teacher education programs have been self-accrediting institutions and are progressively losing autonomy as national policies are being implemented by state-based teacher registration boards. The move to national standards is argued to benefit children and teachers who move between Australian states and territories who now experience common programs, curriculum, and standards.

The second theme is the development of national professional standards which define the wide range of knowledge and skills required for graduating as a teacher, highlighting the complex knowledge base required for teaching. As Shulman (1986) noted, teachers not only require knowledge of students and their learning, subject matter knowledge, and PK, but also a synthesis of these knowledges into PCK. The challenge for ITE is not only developing each of these separate knowledge bases but also providing ways of supporting students to bring them together as PCK through planning learning sequences and practicum experiences. Many reviews of teacher education have been critical about the lack of connection between theory and practice and the weak collaboration between universities and schools who offer practicum placements. This discussion is not helped by uninformed and opinionated political commentary about teachers needing to be "trained" and not acknowledging the importance of (science) education research informing practice (For more details, see Treagust, Won, Petersen, & Wynne, 2015, p. 82–83). Both AITSL and the TEMAG review of ITE in Australia have stressed the importance of strong and sustained partnerships between universities and schools in delivering ITE.

The third theme relates to the above comment in that there are two drivers of educational reform: education research which provides new insights into effective practice and politically initiated shifts in education policy and curriculum framing. These reform agendas require ongoing teacher professional learning to support teachers in developing the new knowledge and skills required to implement the reforms, and also, modifications to ITE programs.

REFERENCES

Ainley, J., Kos, J., & Nicholas, M. (2008). *Participation in science, mathematics and technology in Australian education.* ACER Research Monographs. Retrieved from http://research.acer.edu.au/acer_monographs/4

Alexander, R. J. (2000). *Culture and pedagogy: International comparisons in primary education.* Oxford, England: Blackwell.

Australian Government Productivity Commission (2012). *Schools workforce.* Melbourne, VIC: Commonwealth of Australia. Retrieved from http://www.pc.gov.au/inquiries/completed/education-workforce-schools/report/schools-workforce.pdf

Australian Institute of Teaching and School Leadership. (2011). *National professional standards for teachers.* Carlton South, VIC, Australia: Education Services Australia.

Australian Institute of Teaching and School Leadership. (2013). *Accreditation of initial teacher education programs in Australia: Guide to the accreditation process.* Carlton South, VIC, Australia: Education Services Australia.

Australian Institute of Teaching and School Leadership. (2014). *Initial teacher education: Data report 2014.* Melbourne, Australia: Author.

Australian Science Teachers Association. (2009). *National professional standards for highly accomplished teachers of science.* Deakin, ACT: Author.

Beresford, Q., Partington, G., & Gower, G. (Eds.). (2012). *Reform and resistance in Aboriginal education* (Rev ed.). Perth, WA: UWA Press.

Chigeza, P. (2008). Language negotiations indigenous students navigate when learning science. *The Australian Journal of Indigenous Education, 38,* 91–97.

Ferrari, J., & Rushton, G. (2014, May 21). Alarm bell tolls: Teaching courses a class below. *The Australian,* pp. 1, 6.

Finger, G. (2014, June 10). *What the review of teacher education should be asking.* Retrieved from http://theconversation.com/what-the-review-of-teacher-education-should-be-asking-27066

Gilbert, J. K., & Treagust, D. F. (Eds.). (2009). *Multiple representations in chemical education.* Berlin, Germany: Springer.

Gower, G., & Byrne, M. (2012). Becoming a culturally competent teacher: Beginning the journey. In Q. Beresford & G. Partington (Eds.), *Reform and resistance in Aboriginal education* (2nd ed., pp. 379–402). Perth, Australia: UWA Press.

Hackling, M., Byrne, M., Gower, G., & Anderson, K. (2015). A pedagogical model for engaging Aboriginal children with science learning. *Teaching Science, 61*(1), 27–39.

Hackling, M., & Sherriff, B. (2015). Language-based reasoning in primary science. *Teaching Science, 61*(2), 14–25.

Hackling, M., Chen, S., & Romain, G. (2017). Social and cultural factors framing the teaching and learning of primary science in Australia, Germany, and Taiwan. In M. Hackling, J. Ramseger, & S. Chen (Eds.). *Quality teaching in primary science education* (pp. 19–47). Berlin, Germany: Springer.

Hackling, M., Murcia, K., & Ibrahim-Didi, K. (2013). Teacher orchestration of multimodal resources to support the construction of an explanation in a Year 4 Astronomy topic. *Teaching Science, 59*(1), 7–15.

Harris, K.-L., Jensz, F., & Baldwin, G. (2005). *Who's teaching science? Meeting the demand for qualified science teachers in Australian secondary schools.* Melbourne, Australia: Centre for the Study of Higher Education, The University of Melbourne.

Hattie, J. (2009). *Visible learning: A synthesis of over 800 meta-analyses relating to achievement.* London, England: Routledge.

Hobbs, L. (2013). Teaching "out-of-field" as a boundary-crossing event: Factors shaping teacher identity. *International Journal of Science and Mathematics Education, 11*(2), 271–297.

Houses of Parliament, Parliamentary Office of Science & Technology. (2013, March). *STEM education for 14–19 year olds.* Postnote number 430. Retrieved from http://researchbriefings.files.parliament.uk/documents/POST-PN-430/POST-PN-430.pdf

Jones, B. T. (2014, February 20). *Pyne's review panel: Will it help improve teacher quality?* Retrieved from http://theconversation.com/pynes-review-panel-will-it-improve-teacher-quality-23466

Keengwe, J., & Onchwari, G. (2011). Fostering meaningful student learning through constructivist pedagogy and technology integration. *International Journal of*

Information and Communication Technology Education, 7(4), 1–10. doi:10.4018/jicte.2011100101

Kennedy, J., Lyons, T., & Quinn, F. (2014). The continuing decline of science and mathematics enrolments in Australian high schools. *Teaching Science, 60*(2), 34–46.

Koehler, M. J., & Mishra, P. (2009). What is pedagogical content knowledge? *Contemporary Issues in Technology and Teacher Education, 9*(1), 60–70.

Kress, G., Charalampos, T., Jewitt, C., & Ogborn, J. (2001). *Multimodal teaching and learning: The rhetoric of the science classroom.* London, England: Continuum.

Lee, O. (2004). Teacher change in beliefs and practices in science and literacy instruction with English language learners. *Journal of Research in Science Teaching, 41,* 65–93.

Lemke, J. L. (1990). *Talking science: Language, learning and values.* Norwood, NJ: Ablex.

Louden, W. (2015). *Standardised assessment of initial teacher education: Environmental scan and case studies.* Retrieved from http://www.aitsl.edu.au/docs/default-source/initial-teacher-education-resources/standardised-assessment-of-ite_environmental-scan-and-case-studies.pdf

McDonald, T. (2013). *Classroom management: Engaging students in learning.* Melbourne, Australia: Victoria Oxford University Press.

McKenzie, P., Weldon, P., Rowley, G., Murphy, M., & McMillan J. (2014). *Staff in Australia's schools 2013: Main report on the survey.* Canberra ACT: Australian Council of Educational Research.

Ministerial Council on Education Employment Training and Youth Affairs. (2012). *MCEETYA joint statement on education and training in the information economy.* Canberra, ACT: Ministerial Council on Education, Employment, Training and Youth Affairs.

NSW Department of Education. (2013). *Great teaching, inspired learning: A blueprint for action.* Sydney, Australia: NSW Department of Education.

Office of the Chief Scientist. (2012). *Mathematics, engineering & science in the national interest.* Canberra, ACT: Commonwealth of Australia.

Office of the Chief Scientist. (2012). *Mathematics, engineering and science in the national interest.* Retrieved from http://www.chiefscientist.gov.au/wp-content/uploads/Office-of-the-Chief-Scientist-MES-Report-8-May-2012.pdf

Pechenkina, E., & Anderson, I. (2011). *Indigenous Australian higher education: Trends, initiatives and policy implications.* Canberra, ACT: Panel of the Review of Higher Education Access and Outcomes for Aboriginal and Torres Strait Islander People, Department of Education and Training, Commonwealth of Australia.

PriceWaterhouseCooper. (2015). A *smart move: Future-proofing Australia's workforce by growing skills in science, technology, engineering and maths (STEM).* Retrieved from http://www.pwc.com.au/about-us/stem/index.htm

Pyne, C. (2014, Feb 18). A quality education begins with the best teachers. *The Sydney Morning Herald.* Retrieved from http://www.smh.com.au/action/printArticle?id=5177737

Shulman, L. S. (1986). Those who understand: Knowledge growth in teaching. *Educational Researcher, 15*(2), 4–14.

Smart, J. B., & Marshall, J. C. (2013). Interactions between classroom discourse, teacher questioning, and student cognitive engagement in middle school science. *Journal of Science Teacher Education, 24,* 249–267.

Teacher Education Ministerial Advisory Group (TEMAG). (2015). *Action now: Classroom ready teachers.* Canberra, Australia: Australian Department of Education. Retrieved from https://www.studentsfirst.gov.au/teacher-education-ministerial-advisory-group

The New Work Order: Ensuring Young Australians Have the Skills and Experience for the Jobs of the Future, Not the Past. (2015). Retrieved from http://www.fya.org.au/wp-content/uploads/2015/08/fya-future-of-work-report-final-lr.pdf

Thomson, S., De Bertoli, L., & Buckley, S. (2013). *PISA 2012: How Australia measures up.* Camberwell, VIC: Australian Council for Educational Research.

Treagust, D. F., &, Tsui, C-Y. (Eds.). (2013). *Multiple representations in biological education.* Berlin, Germany: Springer.

Treagust, D. F., Won, M., Petersen, J., & Wynne, G. (2015). Science teacher education in Australia: Initiatives and challenges to improve the quality of teaching. *Journal of Science Teacher Education, 26,* 81–98.

Trewin, D., & Madden, R. (2005). *The health and welfare of Australia's Aboriginal and Torres Strait Islander peoples.* Canberra, ACT: Australian Bureau of Statistics.

Tytler, R., & Aranda, G. (2015). Expert teachers' discursive moves in science classroom interactive talk. *International Journal of Science and Mathematics Education, 13,* 425–446.

White House Office of Science and Technology Policy. (2015, February). *Investing in America's future: Preparing students with STEM skills.* Retrieved from https://www.whitehouse.gov/sites/default/files/microsites/ostp/stem_fact_sheet_2016_budget_0.pdf

CHAPTER 12

SECONDARY SCIENCE TEACHER PREPARATION IN THE PEOPLE'S REPUBLIC OF CHINA

Sudong Pan
East China Normal University

Xiaoting Yue
East China Normal University

The closed, traditional education of China did not train science teachers. The preparation of science teachers developed along with the introduction of Western science into modern style schools, after the door was opened by the Opium War (1840). In 1867, *Jingshi Tongwenguan*—the first official modern style school in China, established in 1862 to train foreign language translators—launched the Department of Astronomy and Mathematics, and was thereby the first school to officially start teaching Western science and technology. In the following decades, as modern style education gradually burgeoned in China, other modern style schools of a similar nature started offering science courses as well. At the beginning, science

Model Science Teacher Preparation Programs, pages 249–269

teachers were mainly foreigners, Chinese overseas students, and graduates of mission schools. With the development of the new education and the growth of society, on a national scale, those teachers were unable to meet the needs of school education. After the foundation of Chinese teacher education system, students with science education orientation began to be trained institutionally, and these students gradually became the mainstay of science teachers.

At present, there are two kinds of science curriculum in secondary schools, the separate and the integrated, and accordingly two types of science teachers need to be trained in secondary science teacher education. For a long time, secondary science curriculum was offered separately in China, while junior and senior high schools offered physics, chemistry, and biology courses for all students. In 2001, the Basic Education Curriculum Reform stated, "separate and integrated courses are offered paralleled in junior high school... integrated courses are advocated" (Ministry of Education, 2001, p. 908). Since then, despite the advocacy of the Ministry of Education of People's Republic of China (hereafter referred to MOE), there has been little positive response in practice. As a result, junior high schools in Zhejiang Province and Shanghai along with a small number of junior high schools in other provinces currently offer the integrated science curriculum, while most of junior high schools offer a separate science curriculum. The requirements for science teachers in terms of knowledge and skills for these two types of curriculum are different, and so are the demands on the preparation of science teachers. In order to distinguish between these two streams of teachers, scholars and practitioners refer to them as separate science teacher and integrated science teacher.

DEVELOPMENT OF CHINESE TEACHER EDUCATION

Nanyang Public School founded the first teacher school in 1897—initially enrolling students with primary and secondary education orientation—which marked the beginning of the Chinese teacher education. Chinese teacher education, which has been in place for more than 100 years, can broadly be divided into three periods: independent, closed, and opened teacher education systems.

Period of Independent Teacher Education System

Early Chinese teacher education was developed in imitation of teacher education systems of foreign countries, mainly Japan, but also drawing from teacher training systems in Germany, France, and the United States.

In order to promote the institutionalization of modern style education, the Qing Dynasty (1636–1912) government imitated Japan's teacher education system, and began to establish the Chinese teacher education system. According to the junior teacher school charter and the senior teacher school charter in the school system of 1904, junior teacher schools trained teachers for primary schools, while senior teacher schools trained teachers for high schools and junior teacher schools.

Along with the arduous imitations was a series of domestic and international political and social turbulences such as the civil wars, and the Anti-Japanese war (1938–1945), which made the process of constructing the educational system extremely hard. In spite of these hardships, the government of the Republic of China (ROC) established a relatively independent two-tier public teacher education system: teacher schools (senior high school level training primary school teachers), and teacher colleges (higher education level training secondary school teachers). "In underdeveloped economic and educational conditions, independent teacher education system could quickly foster qualified and needed teachers" (Ma, 2003, p. 29).

Period of Closed Teacher Education System

After the establishment of the People's Republic of China (hereafter referred to as China) in 1949, China imitated the Soviet Union comprehensively, establishing a socialist planning society system. China followed the teacher education system of the Soviet Union, and attempted to establish a closed independent teacher education system, adapted to the centralized planning society. First came the reconstruction of teacher education institutions—in particular, the adjustment of higher education in the 1950s—setting up a group of teacher colleges or universities to prepare and train secondary school teachers. Second, the *Sovietization* of educational theory was vastly taken up—in particular, the far-reaching *Kairov Educational theory*—which deeply impacted the theory and practice of China's teacher education (Jin, 2008, p. 69). Third, influenced by the educational experience of the Soviet Union, China commenced a widespread teaching reform, ranging from changes in the training programs, curriculum, and course syllabuses, to the change of instructional method and evaluation. The Chinese government gradually built a three-tier teacher education apparatus (Table 12.1). It should be noted that teacher education and students' tuition was free, and certain financial aid and professional scholarships were granted. The enrollment and graduate destination of student teachers was *targeted enrollment and training,*[1] and student teachers after graduation were assigned to schools or kindergartens as teachers. Until the 1990s, except

TABLE 12.1 Chinese Three-Tier Teacher Education Apparatus (1950s–1990s)

	Teacher Education Institution	Total Duration	Enrollment Target	Teacher of After Graduation
First Tier	Teacher school	3–4 years	Junior high school graduate	Primary schools and early childhood education
Second Tier	Teacher junior colleges	2–3 years	Senior high school graduates	Junior high school teacher
Third Tier	Teacher college or university	4 years	Senior high school graduates	Senior high school teachers or secondary schools teachers

Source: Ma, 2003, p.195

the catastrophic Cultural Revolution (1966–1976),[2] China had been using this closed and targeted teacher education system.

This teacher education system matched the planned economy system, mainly based on the teacher education system of the Soviet Union, but also factoring in the social and economic considerations—such as, more population, a weaker economic foundation—in China. However, along with the immense social development, there increasingly arose a need for more educated people. As such, the government greatly expanded the size of primary and secondary education programs, but was unable to provide commensurate educational funds, below 4% of GDP over a long time, which is perhaps why it was referred to as a "poor country running big education." Despite being under-funded, the system managed to train as many teachers as possible, guarantee to assign a certain number of new teachers to a certain province or region every year, and to ensure planned education development.

Period of Opened Teacher Education System

The Chinese Economic Reform (1978–present) gradually broke free from the rigid, planned economy system, as China witnessed increasing social, political, and economic development in the 1990s, which ushered in fundamental changes in the teacher education system, leading to the the cessation of the (original) closed teacher education system, and thereby the emergence of an opened teacher education system. In 1992, China planned to build its *Socialist Market Economic Society System,* and the government began to "embrace a sweeping wave of neo-liberal ideology, e.g., marketization, privatization, and decentralization" (Li, 2012, p. 422). Accordingly, primary and secondary education changed from quantity-oriented to quality-oriented, and from stiff to flexible, which accompanied considerable changes in teacher education. In this regard, three significant changes came to pass. Firstly, teacher colleges and universities were encouraged to launch non-teacher education orientation programs and to train students of non-teacher education orientation. Secondly, as stated by the Central Committee of Communist Party of China and State Council in 1999, "universities and colleges of non-teacher education orientation participating in fostering primary and secondary school teachers, and qualified universities exploring to run teacher education schools are encouraged" (as cited in He, 2003, p. 289). To host such programs, colleges and universities of non-teacher education orientation were also authorized to develop teacher education programs and train student teachers. Lastly, with the tuition policy change over time, students attending teacher preparation institutions began to be be required to pay full tuition fees in 2000. The free teacher education model, which existed for more than 90 years, was replaced by the paid teacher education model. In this way, the gap between teacher education and non-teacher orientation education was in a sense bridged, and an opened and flexible teacher education system suitable for the socialist market economic society system was constructed.

The original three-tier teacher education apparatus has now evolved into a two-tier one with the development of teacher education and the improvement of primary and secondary teacher education requirements. Teacher education institutions prepared and trained an increasing number of student teachers, particularly after the higher education expansion (1999–2012), when a large number of student teachers with bachelor's degrees were graduated. China's rapid social development increasingly called for stricter requirements to be placed on the quality of primary and secondary school teacher education programs. More specifically, the *Education Revitalization Action Plan for the 21st Century* (MOE, 1999) called on "around 2010, in general region, to strive to make primary and junior high school teachers's qualification promoted to junior college[3] level and undergraduate-level respectively, in developed area, high school teachers and principals in the master's degree should reach a certain percentage" (p. 218). The graduates of teacher schools, equivalent to the high school level, could no longer meet the need of primary schools or kindergartens. These schools have been either upgraded to junior college or merged, or canceled. The graduates of junior teacher colleges can not also meet the need of junior high schools, since these junior colleges have been either upgraded to teacher colleges, or merged into other teacher colleges or universities. Thus, the Chinese teacher education apparatus became two-tier, replacing the original three-tier one, as shown in Table 12.2. There are 62 junior teacher colleges, or first-tier institutions, which train primary or kindergarten teachers, while there are 116 teacher colleges and universities, or second-tier institution, which prepare and train

TABLE 12.2 Chinese Two-Tier Teacher Education Apparatus (At Present)

	Teacher Train Institutions	Total Duration	Enrollment Target	Teacher of After Graduation
First Tier	Teacher junior colleges	5 years	Junior high school graduate	Primary schools and kindergarten
		3 years	Senior high school graduate	
Second Tier	Teacher college or university	4 years	Senior high school graduate	Primary and secondary schools, and kindergarten

primary and secondary school, or kindergarten teachers (MOE, 2015b). In 1997, China began to enroll in-service teacher studying for Master of Education (MEd), and as many as 83 higher education institutions have been approved and qualified for training MEd students (Yang & Zhao, 2013). The average educational level of primary and secondary school teachers has gradually increased, and the teacher education apparatus may develop a new three-tier one in the future, including junior teacher college, teacher college, and postgraduate education institution of MEd.

STANDARDS FOR SCIENCE TEACHER EDUCATION AND NATIONAL TEACHER QUALIFICATION CERTIFICATION FOR SCIENCE TEACHERS

Standards for Science Teacher Education

After the Cultural Revolution (1966–1976), while the teacher education system before the Cultural Revolution was recovered, there were no clear curriculum standards for science teacher education in China. Mainly based on the major norms issued by the College and University Teaching Guide Committee of MOE, a major syllabus of science education, which included a variety of mandates, was developed. For example, the physics education major established its major syllabus by means of the syllabus of physics major, and the chemistry education major established its major syllabus by means of the syllabus of chemistry major, and so on. As such, SMK (subject matter knowledge) courses were attached great importance, and GPK (general pedagogical knowledge) and PCK (pedagogical content knowledge) courses were attached less importance. GPK and PCK courses only contained the "traditional threes": pedagogy, psychology and subject didactics. In an attempt to gain more educational resources, some higher teacher education institutions strived to heighten their academic research level, while weakening their teacher education programs—and also attending less to

GPK and PCK courses—which resulted in some defects in student teachers' knowledge and ability structure.

In order to reverse the unfavorable situation, and to effectively improve the quality of science teachers training, MOE committed to development of teacher education standards in 2010s. The *Teacher Education Curriculum Standard* (trial, 2011) and the *Teacher Professional Standards* (trial, 2012) were issued. The latter further includes three sub-standards:

- Kindergarten Teacher's Professional Standard (trial)
- Primary School Teacher's Professional Standard (trial)
- Secondary School Teacher's Professional Standard (trial)

Enacting the aforementioned standards demands a new set of fundamental rules related to educational beliefs and professional contents, a new direction for (science) teacher education reform, all of which point to an important progress in the construction of (new) science teacher education system in China. According to the *Secondary School Teacher's Professional Standard* (trial; hereafter referred to Standard), a (science) teacher must have four basic educational beliefs:

- focusing on teaching ethics
- basing on all of students
- grounding in ability
- lifelong learning

His/her professional basic contents consist of 3 dimensions: professional philosophy and ethics, professional knowledge, and professional ability (see Table 12.3). These tenets relate to the GE (general education), SMK (subject matter knowledge), GPK (general pedagogical knowledge), and PCK (pedagogical content knowledge) courses of teacher education.

National Teacher Qualification Certification Rules for Science Teacher

For a long time, China had not yet set up a (science) teacher qualification certificate rule, and a candidate obtaining the teacher qualification or not primarily depends on his/her diploma. After the Cultural Revolution, the then-faulty society operation mechanism was restored, and although the (severely damaged) primary and secondary education system was rebuilt, there seemed to be a severe shortage for secondary science teachers with qualified educational background. When a science teacher was hired, his/her educational background was regarded as the most important criterion.

TABLE 12.3 Secondary School Teachers' Professional Basic Contents

Dimensions	Concrete Contents
Professional Philosophy and Ethics	• career understanding and awareness • attitude and behavior on students • attitude and behavior on education and teaching • personal cultivation and behavior
Professional Knowledge	• general pedagogical knowledge • subject matter knowledge • subject teaching knowledge • general knowledge
Professional Abilities	• instructional design • instructional implementation • class management and education activities • educational and instructional evaluation • communication and cooperation • ability of reflection and professional development

Source: MOE, 2012

As long as a science student teacher graduated from a teacher institution, he/she would directly gain the science teacher qualification.

In the *Teacher Law*, enacted in 1993, the establishment of a (science) teacher qualification rule was formally proposed for the first time. The law clearly specified the minimum educational background of a school science teacher at any level, but no specific requirements for knowledge and abilities, particularly educational knowledge and abilities, were dictated (see Table 12.4).

A *double-track* science teacher qualification certification examination was subsequently introduced in China, which differentiated graduates of science education major and graduates of science major. The *Teacher Qualification Rule* (State Council, 1995) clearly defined the (science) teacher qualification for the graduate of a non-oriented teacher education institution or a citizen in society. In 1998, the science teacher qualification certification examination for a non-oriented teacher institution graduate, or citizens was piloted in six provinces. The paper test of science teacher qualification, along

TABLE 12.4 Regulations of Teacher Qualification in the Teacher Law

Teacher of Different Level School	Regulations of Qualification
Kindergarten teacher	Graduated from a teacher school of early childhood education and above
Primary school teacher	Graduated from a teacher school and above
Junior high school teacher	Graduated from a junior teacher college and above
Senior high school teacher	Graduated from a teacher college or university and above

Source: National People's Congress,1993

with the organization of examinations, was administered by each province respectively. The implementing regulation of the teacher qualification rule, enacted subsequently, stated that a graduate of teacher school or college was qualified, provided that he/she obtained a satisfactory level of Mandarin. This status, along with good physical and mental qualities, could allow for the obtainment of teacher qualification certification (MOE, 2000).

Starting in 2015, the unified national (science) teacher qualification certification regulation is established, which offers a *single-track* both for student teachers who have majored in science education and students having majored in science. According to the policy document of the MOE, a teacher education institution graduate can no longer directly gain the teacher qualification certification. Instead, these candidates are gradually merged into the track of the teacher qualification certification exam. This model is first implemented in some provinces, and then will gradually be applied to all provinces in China. Similar to graduates from institutions of non-teacher education orientation, student teachers who have majored in science education cannot directly attain teacher qualification certification, unless they pass the (teacher) qualification certification exam.

Secondary teacher qualification certification is divided into two levels: junior high school and senior high school levels. The former includes 15 kinds of subject teachers—such as physics, chemistry, biology, and integrated science teachers—whereas, the latter entails 14 kinds of subject teachers—like physics, chemistry, biology, and technology.

The national unified Teacher Qualification Certification Examination (hereafter referred to as the Examination), encompasses theoretical and practical assessments, and is intended to comprehensively evaluate a teacher candidate's professional quality, by using a variety of assessment methods. The Examination is composed of a paper-and-pencil test as well as an interview session. The paper-and-pencil test contains three sub-tests, as shown in Table 12.5. These tests are more related to GPK and PCK, and less related to GE and SMK. The three (sub-test) papers of science teacher qualification examination are administered by the Examination Center of MOE.

A typical interview session examines a candidate through instructional design (or activity design), simulation lesson (or demonstration), question-and-answer—which adopts the mode of structured interview—and situational simulation, etc. The content of the interview session (mainly related to PCK and GPK) includes

- professional awareness,
- psychological quality,
- instructional manners,
- verbal expression,
- quality of thinking,

TABLE 12.5 Paper-and-Pencil Tests of the Teacher Qualification Certification Examination

Sub-Test	Content Modules
Comprehensive Quality of Secondary School	• Professional beliefs • Education laws and regulations • Professional ethics for teachers • Cultural literacy • Basic abilities
Educational Knowledge and Ability of Secondary School	• Basics and fundamentals of pedagogy • Secondary curriculum • Secondary teaching • Students' learning psychology • Secondary students' psychological development • Secondary student mental health education • Secondary student moral education, class management • Teachers' psychology of secondary school
Subject Knowledge and Teaching Ability	• Subject knowledge and pedagogical content knowledge • Instructional design • Instructional implementation • Instructional evaluation

- instructional design,
- instructional implementation, and
- and instructional evaluation.

When a candidate successfully passes all of the paper-and-pencil tests, he/she can take part in the interview session. Once the candidate has passed both the paper-and-pencil test and the interview session, he/she can get the qualification certification.

The Standard and the Examination have a high internal consistency. The contents of the Examination are related to the three dimensions of the Standard, and the basic ideas of the Examination are consistent with the Standard. The introduction of the Standard and the Examination in the 2010s reflects the notable change in the Chinese teacher education, particularly in terms of quality. The system has seen a shift from the (incomplete and inadequate) quality view of teacher education emphasizing teachers' subject knowledge and capacity in 1980s, to evolve into present comprehensive quality view of teacher education emphasizing teachers' professional quality.

SYLLABUS OF SCIENCE TEACHER PREPARATION IN CHINA

Considering that Chinese science teachers are divided into separate science teachers and integrated science teachers, their training syllabuses are different. These respective traits will be elaborated in the following sections.

Separate Science Teacher

A separate secondary science teacher is trained in a specialized department of a teacher college or university, such as department of physics, chemistry, and biology. As an example, let's take the major syllabus of physics education at East China Normal University (abbreviated as ECNU), which prepares and trains "separate" science teachers. According to the syllabus, a student teacher needs to complete 155 credit hours for graduation (ECNU, 2013). The following are the introduction of the SMK, GPK, and PCK courses in the syllabus.

SMK (Subject Matter Knowledge) Courses

There are 30 obligatory SMK courses at ECNU, including fundamental physics, and experiments, theoretical physics, and experiments, advanced mathematics, linear algebra, statistics, methods of mathematical physics, electronics, physics history and methodology, graduation theses, and so on. Both the obligatory and specified elective SMK courses accumulate 77.5 credit hours, accounting for 50% of the total credit hours, which reflects the notion that China attaches special importance to the training of SMK and subject mastery of science teachers. Six SMK courses are directly related to the Examination, accounting for a total of 17 credit hours (ECNU, 2013). The instructional method of the courses is lecture oriented, plus a small amount of question-and-answer interaction, while instructional assessment is mainly a closed, paper-and-pencil test.

GPK (General Pedagogical Knowledge) Courses

GPK courses at ECNU not only occupy few credit hours, but the content is only an introduction to the basic educational theory, psychology theory, and educational technology knowledge. Compulsory and elective GPK courses have 10 credit hours, which represents 6.5% of the total credit hours. There are four compulsory courses: pedagogy, psychology, teacher spoken language (mainly *Mandarin* training), and educational technology. In addition, about 50 teacher professional quality series courses are available, such as instructional design, learning theory, educational psychology, classroom management, research methods, and teachers' professional development, professional ethics of teachers, and so on, basically involving the professional knowledge and skills required by the Standard. However, due to the excessive number of courses and content dispersion, there is a small likelihood for each course to be elected, thereby the practical result of these courses differs from the original intent.

PCK (Pedagogical Content Knowledge) Courses

Compared with SMK courses, the credit hours of PCK courses at ECNU are less, counting up to 18.75, which represents 12.1% of the total credit hours. PCK courses include seven compulsory courses: secondary physics instructional theory, secondary physics instructional design, secondary physics instructional skills training, instructional skills training of secondary physics experiment, micro-teaching, educational practicum I and II. They also include two elective courses: instructional evaluation of secondary physics, and multimedia courseware design of secondary physics. In addition, among the SMK courses, only physics history and methodology is concerned with PCK.

As with SMK courses, the instructional method of the theory courses of PCK is mainly lecture-oriented, in addition to some case-analysis, while instructional assessment is mainly closed paper-and-pencil tests. Skills training, microteaching, and other practical courses of PCK are completed in teams with the guidance of teachers at ECNU, and assessment is completed through performance assessment. Educational practicum I (short field visit) is a group of student teachers, who observe teaching and learning and extracurricular activities in a high school classroom. Student teachers also attend the classes of secondary science experienced teachers, observing the class management and learning the guidelines of extracurricular activities of science teachers in order to gain practical knowledge and field experiences. Assessment of practicum I is based on the quality of a variety of observation reports. In educational practicum II, which lasts 12 weeks, student teachers are sent to high schools to get rich educational field experience. The instructional method of the educational practicum II is "apprenticeship-style," and an experienced science teacher of high school classes gives special guidance to 1 or 2 interns. In the first stage, he/she will let interns observe his/her lessons, mark the homework, and then guide them to write instructional design, and so on. In the second stage, the interns will try to deliver some lessons (at least 5 new lessons) to the secondary students, under the guidance of secondary science teacher and a university faculty member. Assessment of educational practicum II is also based on their performance.

Integrated Science Teacher

There are 65 colleges or universities in China that offer the integrated science education program (Ding, 2015). The programs are usually hosted by departments of physics, chemistry and biology, and other institutions as well. These institutions prepare and train integrated science teachers and separate science teachers simultaneously. At Hangzhou Normal University (abbreviated as HZNU), integrated science teachers are trained in department of physics. According to the syllabus of integrated science education, a student teacher needs to complete 168 credit hours for graduation (HZNU, 2012).

SMK Courses

The credit hours for the SMK courses of the integrated science education major at HZNU is made up of 62 points, accounting for 44.6% of the total credit hours. The SMK courses mainly contain advanced mathematics, fundamental physics, chemistry, life sciences, astronomy, geological science, and physics, chemistry, life science experiment, and electronics, scientific thinking and methodology, introduction of HPS, and graduation thesis (HZNU, 2012). The instructional method and assessment of these courses, as well as those of the following GPK and PCK courses are the same as those at ECNU.

GPK Courses

The GPK courses at HZNU count up to 11 credit hours, and occupy 6.5% of the total credit hours. Although the credit hours of GPK are less, there are more GPK courses, and the courses involve a wide spectrum of content. Seven GPK courses are offered: basic pedagogy, basic psychology, teacher skills training, professional ethics of teachers, classroom management skills, educational research methods, and modern educational technology.

PCK Courses

The PCK courses at HZNU account for 22 credit hours, and occupy 13.1% of the total credit hours. Six PCK courses are offered: science instructional theory, basic science education research, secondary science experiment, psychology of science learning, and educational practicum I and II. Scientific thinking and methodology, introduction of HPS of SMK courses are concerned with PCK.

Comparison of Two Science Teacher Programs

The syllabuses of ECNU and HZNU show that the programs of separate science teachers and integrated science teachers have more similarities than differences. First, the proportions of the three types of courses are basically the same. SMK courses constitute close to 50% of the total credit hours, while PCK and GPK respectively account for nearly 12% and 6% of the total credit hours, indicating the differential importance attached to these courses, with SMK receiving most attention, PCK less, and GPK being the least attended to course. Second, instructional modes, strategies, and assessment of the two programs are basically the same, and have changed little in comparison with the past, for example, theory courses are still mainly traditional lecture-oriented. Third, practical courses seem to have changed more than the theoretical ones, for instance educational practicum has been extended into two stages, a short and a long time component. Fourth, the requirements in terms of knowledge and abilities in SMK courses vary greatly. SMK of a separate teacher stream is profound and deep, while that of an integrated teacher is wide and shallow. In addition, there are also some differences in GPK and PCK courses, for example regarding course content and credit hours.

Three types of courses of the two syllabuses are referred to as the Examination, but correlative extent of the Examination is different. PCK and GPK courses have the highest correlative value to the Examination, both are closely related to the interview session, and related to the sub-test of the Physics (or Science) Subject Knowledge and Teaching Ability, and the Education Knowledge and Ability, respectively. These two types of courses account for a low percentage in the syllabus, reflecting low attention to those. SMK courses receive most attention at ECNU and HZNU, but involve a small amount in sub-test of the Physics (or Science) Subject Knowledge and Teaching Ability.

DISCUSSION

The Advantages of Science Teacher Preparation in China

Establishment of a Science Teacher Preparation System Suitable to the Social Development

In each period of social transition, China tended to imitate foreign teacher education system, then made adjustments based on its national condition and educational development, in an attempt to make the (science) teacher education system suitable for social development. After independent, and closed science teacher education system, an opened science teacher education system has been established and is still in place until now. This system is adapted to cater to decentralization, flexibility, diversification, and marketization that the socialist market economic society system espoused. The system can simultaneously prepare and train separate science teachers qualified for secondary separate science curriculum, and integrated science teachers qualified for integrated sciences curriculum in junior high schools of some regions. At present, there are 43.84 million junior high school students, 24.00 million high school students, 3.48 million junior high school teachers, and 1.66 million high school teachers (MOE, 2015a). There are estimated 1.25–1.30 million secondary science teachers, and a majority of secondary science teachers have been trained by the system. Those teachers are in charge of teaching science to (nearly 70 million) high school students, and have thus made an important contribution to the development of science education and the social development.

Standardizing Secondary Science Teacher Preparation by Laws and Policy Documents

China has promulgated the *Teacher Law,* the *Teacher Qualification Rule,* and *Implementing Regulations of the Teacher Qualification Rule* since 1990s, and those relevant to the teacher qualification certification were constructed step by step. From law to rule, and regulation, the rules related to science teacher qualification certification become complete. In addition, each law or rule was first implemented in one or several provinces, and then applied to the whole country, after a period of modifications. By doing this, the negative effects of urgent implementation of a law or regulation can be avoided.

Combination of laws or regulations, and policy documents not only provides the basis for the specification of science teacher preparation, but also points out the reformation direction for science teacher preparation programs and curriculum in teacher colleges and universities. In light of China's specific national conditions—for instance, the leadership of Central Committee of Communist Party of China (CCCP), socialist society system—besides laws, rules and regulations, the policy documents of CCCP, State Council, and other administration departments also play an important role. Some of these important policy documents include the *Education Revitalization Action Plan for the 21st Century* (MOE, 1998), *Decision on Deepening Education Reform and Comprehensively Promoting Quality Education* (He, 2003), *Basic Education Curriculum Reform (trial)* (MOE, 2001), and other policy documents issued by CCCP, or State Council, or MOE, and also have played a significant role in secondary science teacher preparation.

National Science Teacher Qualification Certification Examination and the Standard *Leading the Science Teacher Preparation Curriculum Reform*

The unified science teacher qualification certification examination on a national scale improves the teachers' entry threshold, and tends to heighten the quality of teachers to a certain extent. The original "dual-track" teacher qualification certification examination allowed some barely graduated student teachers to obtain teacher qualification certification.[4] While "single-track" teacher qualification certification examination filters out some student teachers, not allowing those with poor knowledge bases and abilities to obtain teacher qualification certifications, it helps to strengthen the quality of science teacher preparation.

The Standard can standardize the training process of science teachers, while the Examination can examine the quality of science teachers training, and the combination of both has led to the curriculum reformation of science education in teacher institutions in past years. Firstly, GPK courses have undergone change, for example, in addition to adding to the number of GPK courses, the content of traditional pedagogy and psychology courses has also changed. Secondly, PCK courses are also changed; that is, both the credit hours and the content of PCK courses have increased, extending to a range of courses from traditional subject pedagogy, and increasing the coverage of the content. Thirdly, educational practicum is changed; that is, the time span of educational practicum has increased from the original 4–6 weeks to 12–16 weeks (from 2 or 3 credit hours to 6 or 8 credit hours); the field course has been expanded into educational practicum I and II, allowing for access to more field education experiences by stages; and especially in some teacher colleges and universities, *Replacement of In-Service Teacher Practicum (Ding Guang Internship* in Chinese)[5] of six months is launched. Due to the reformation of teacher education courses, especially as a direct influence of the Examination, compared with the past, a given science student teacher pays more attention to the learning of PCK and GPK courses, with more emphasis on both theory learning and practical training.

The Shortage of Science Teacher Preparation in China

The Gap Between Science Education Syllabus and the Requirement of the Standard

There is a certain gap between the syllabus of science teacher preparation, and the basic idea and professional content of the *Professional Standards for Secondary School Teacher* (trial). The main features of the syllabuses follows.

Firstly, the syllabuses attach great importance to SMK courses, yet the proportion of credit hours and the number of courses are still too large. The total number of credit hours of SMK courses at ECNU and at HZNU account for more than 40% of the total credit hours, especially near 50% at ECNU. It relates to a certain extent with the Chinese education tradition of "teaching knowledge" and the tendency of "examination-oriented education" (*Ying Shi education,* in Chinese)[6]. When a high school recruits a science teacher, especially for well-known schools, the candidates are screened with more attention to their scientific knowledge bases. The relatively large percentage of the SMK courses enables student teachers to obtain a solid subject foundation, as the total hours of restriction will reduce their learning opportunities of the GPK and PCK courses.

Secondly, the GPK courses are underestimated in the syllabuses. The credit hours of GPK courses at ECNU and HZNU account for 6.5% of total credit hours in the programs. This is much less than the major syllabus of science education at Hiroshima University in Japan, which has 12 obligatory GPK courses in the science education major, and accounts for 18% of total credit hours (Hiroshima University, 2012). The underestimated GPK leads to a lack of at least part of the content required by the Standard, such as, the courses of classroom management, reflection, and development capacity lacked at ECNU, and the courses of instructional evaluation, reflection, and development capacity not available at HZNU.

Lastly, the PCK courses have some defects. PCK credit hours at ECNU and HZNU account for 13% of the total credit hours. Although it is higher than the percentage of GPK courses credit hours, compared to Hiroshima University whose PCK credit hours of science education major account for 21.6% (Hiroshima University, 2012), the gap is relatively large. According to Schneider and Plasman (2011), PCK of science teachers include 5 components and 18 categories, see Table 12.6. Compared to the PCK model, although the syllabus of ECNU has all of the five components, some categories in component 1 and component 2 are lacking. Component 5 is lacking at HZNU, while many categories in the component 1 are missing. The PCKs at both universities are "less comprehensive type." In addition, only one SMK course at ECNU and two SMK courses at HZNU are connected with PCK, which is difficult to reflect that PCK is the "alloy" of SMK and GPK. Those are "weak correlation" PCK, which would undermine the effect of PCK training.

In order to reach the requirements of the Standard, teacher colleges or universities need to increase the credit hours of GPK and PCK courses, and to modify the content of the PCK courses. Due to the unique Chinese higher education policy, political courses must achieve 14 credit hours, comprising nearly 10% of total credit hours, thereby making it difficult to reform GPK and PCK courses.

Problems in Science Teacher Qualification Certification Examination

With the gradual introduction and development of the national science teacher qualification certification examination, it may lead to two more serious problems:

First, there appears the tendency of the "examination-oriented" in some science student teachers. While the Examination requires that students learn a certain amount of content, many learning opportunities are left out of the requirements. Because the GE, SMK, GPK, and PCK courses in the syllabus of science education are not in the same degree of correlation with the Examination, some science student teachers may only be concerned with the courses directly related with the Examination, such as PCK and GPK courses. They may underestimate SMK and GE courses, most of those will not be tested in the Examination, which leads to the flawed structure of their science knowledge and abilities.

TABLE 12.6 Science Teacher PCK, Aspects and Categories

Components of Science Teacher PCK	Categories for Each Component of PCK
Orientations to Teaching Science	Teachers' ideas about... • Purposes and goals for teaching science • The nature of science • The nature of teaching and learning science for students
Student Thinking About Science	Teachers' ideas about... • Students' initial science ideas and experiences (including misconceptions) • Development of science ideas (including process and sequence) • How students express science ideas (including demonstration of understanding, questions, and responses) • Challenging science ideas for students • Appropriate level of science understanding
Instructional Strategies in Science	Teachers' ideas about... • Inquiry strategies (e.g., questions and including how to use, how science is developed, and how student thinking is supported) • Science phenomena strategies (e.g., demonstrations or predict-observe-explain and including how to use, how science presented, how student thinking is supported) • Discourse strategies in science (e.g., argument, writing, presenting, or conferencing and including how to use, how science portrayed, and how student thinking is supported) • General student-centered strategies for science (vs. teacher-centered) including how to use and when, how science is represented, and match to student needs and thinking
Science Curriculum	Teachers' ideas about... • Scope of science (importance of science topics and what science is worth knowing or teaching) • Sequence of science (organizing science content for learning) • Curricular resources available for science • Using standards to guide planning and teaching science
Assessment of Students' Science Learning	Teachers' ideas about... • Strategies for assessing student thinking in science • How or when to use science assessments

Source: Schneider & Plasman, 2011

Second, some students with non-science education orientation pass the Examination through training for "false education teaching abilities." This means the abilities are "shaped" by "specially trained" mode at social training companies. Students obtain these "abilities" to pass the Examination, without the need to participate in educational practicum, and they are only able to answer the questions of the tests in the Examination, but cannot solve the real educational and instructional problems (Wang, 2015). This approach will allow some students who do not have the science teacher qualification to gain certifications despite their lack of educational field experience. Thus, this could undermine the uniqueness of Chinese (science) teacher education system, and even lead to its marginalization. To solve these problems, laws and regulations or policies need to be designed to eliminate or mitigate the negative effects of the Examination.

Teacher Education Curriculum Reformation Needs to be Deepened

In recent years, (science) curriculum reformation in teacher colleges or universities has advanced steadily, especially after the enactment of the national science teacher qualification certification examination, and the Standard. Main reformation practices are to change training syllabuses, for instance, they may need to increase or decrease the number of courses, class hours, or to change the course content. However, the curriculum implementation has been changed slightly.

First, in terms of instructional strategies, GPK and PCK courses mainly use lecture-oriented strategies, and have only increased some practical cases and a few of questions-and-answer activities, allowing for some—but not enough—discussion between teacher trainers and student teachers, or students and students, which are still basically reflective of teacher-centered teaching strategies. Instructional strategies of students-centered learning—such as inquiry-based learning, seminar style classes, and problem based learning (PBL)—are still lacking. Also, the case-based instructional method is still under-utilized in classroom teaching.

Second, in instructional assessment aspects, except for the practical courses of PCK, the instructional assessment of the GPK and PCK courses is still based upon the traditional paper-pencil test. The paper-pencil test makes it difficult to assess the abilities of students, and tends to encourage rote learning, rather than authentic experience. A better option might be a combination of paper-pencil test and alternative assessment, such as portfolio, or performance assessment, etc.

SUMMARY

For more than 100 years, trapped between internal conflict and external interference, several generations of Chinese have been making unremitting efforts, under the two interwoven development models of exogenous and endogenous origin (Jin, 2008, p. 71). The Chinese have managed to successfully build an opened and flexible secondary science teacher preparation system, which adapts to the Chinese socialist market economy system. The teacher preparation system can simultaneously train separate science teachers and integrated science teachers, which hold up the secondary school science education with the most educated population in the world.

With the promulgation of all relevant laws and regulations, and policy documents—particularly, the enactment and implementation of the *Secondary School Teachers Professional Standards* (trial) and the national *(Science) Teacher Qualification Certification Examination*—the curriculum reformation in teacher colleges and universities has gathered more steam, leading to to increase in the normalization degree of science teacher preparation process, and enhancing the quality of science teacher preparation. However, if there are no further supporting regulations or policies, some problems in science teacher preparation—such as the tendency to remain "examination-oriented"—will not be avoided, which could undermine the uniqueness of Chinese (science) teachers education system.

NOTES

1. Targeted enrollment and training is a special approach to teacher education institutions in China. An institution, based on the plan, enrolls a certain number of students in a province each year, and these students must be returned to the province to become teachers after graduation, so as to ensure that the province is under planning for the development of education.
2. The Cultural Revolution (1966–1976) was a catastrophe that brought disaster to the Chinese. The operating mechanism of the society was badly damaged, and the whole country was in the midst of civil strife. Education system as a whole, including the teacher education system, was severely damaged, and many teachers were persecuted, causing great losses to education.
3. Junior college (2–3 years duration) belongs to higher education in China, lower than college and university (4 years gain bachelor's degree).
4. Different to the universities of Western countries, students in China's universities is basically "hard to enter, easy to graduate." Their entrance examinations are very strict, especially for high level universities. But once admitted to the universities, most students will graduate on time, and only a few students fail to graduate or graduate on time. This results in some students who have bad academic knowledge and abilities getting a certificate of graduation through several "course retaking" or "make-up examinations."
5. Replacement of In-Service Teacher Practicum is a new form of educational practicum that is encouraged by the Ministry of Education. A student teacher goes to a bad-conditioned primary or secondary school, fully carries out all duties of the in-service teacher in the school, replaces all educational works of the teacher, allowing the teacher to be trained full time in teacher colleges or universities.
6. It is an educational mode which uses a mechanical way and rote learning to educate students, and whose principal purpose is to cater to the entrance examination. It has affected Chinese basic education for a long time and has become one of the most serious problems of Chinese basic education.

REFERENCES

Ding, B. (2015). Science Teacher Education in Mainland China. In Richard Gunstone (Ed.), *Encyclopedia of Science Education* (pp. 917–924). Dordrecht, The Netherlands: Springer.

East China Normal University. (2013). *The major syllabus of physics education.* Sahnghai, China: Author.

Hangzhou Normal University. (2012). *The major syllabus of science education.* Hangzhou, China: Author.

He, D. (Ed.). (2003). *Important educational laws, regulations and policy documents of People's Republic of China (1998–2002).* Haikou: Hainan Press. (in Chinese).

Hiroshima University. (2012). *The major syllabus of science education.* Hiroshima, Japan: Author.

Jin, Z. (2008). *The history, theory and practice of teacher education.* Shanghai, China: Shanghai Educational Publishing House. (In Chinese).

Ma, X. (2003). *Teacher education history in China.* Beijing, China: Capital Normal University Press. (in Chinese).

Ministry of Education. (1999). Education revitalization action plan for the 21st century. In D. He (Ed., 2003), *Important educational laws, regulations and policy documents of People's Republic of China (1998–2002)* (pp. 217–222). Haikou, China: Hainan Press. (in Chinese).

Ministry of Education. (2000). Implementing regulationof the teacher qualification rule. In D. He (Ed., 2003), *Important educational laws, regulations and policy documents of People's Republic of China (1998–2002)* (pp. 703–704). Haikou, China: Hainan Press. (in Chinese).

Ministry of Education. (2001). *Outline of basic education curriculum reform (trial).* In D. He (Ed., 2003), *Important educational laws, regulations and policy documents of People's Republic of China (1998–2002)* (pp. 907–909). Haikou, China: Hainan Press. (in Chinese).

Ministry of Education. (2015a). *National education development statistical bulletin of 2014.* Retrieved from http://www.moe.gov.cn/jyb_xwfb/gzdt_gzdt/s5987/201507/t20150730_196698.html

Ministry of Education. (2015b). *The list of colleges and universities.* Retrieved from http://www.moe.gov.cn/srcsite/A03/moe_634/201505/t20150521_189479.html

National People's Congress. (1993). Teacher law. In D. He (Ed., 1998). *Important educational laws, regulations and policy documents of People's Republic of China (1949–1997)* (pp. 3570–3572). Haikou, China: Hainan Press. (in Chinese).

Schneider, R. M., & Plasman, K. (2011). Science teacher learning progressions: A review of science teachers' pedagogical content knowledge development. *Review of Educational Research,* 81(4), 530–565.

State Council. (1995). Teacher qualification rule. In D. He (Ed., 1998). *Important educational laws, regulations and policy documents of People's Republic of China (1949–1997)* (pp. 3907–3908). Haikou, China: Hainan Press. (in Chinese).

Wang, J.(2015). National teacher qualification examination on effect of pre-service teacher education. *Educational Management in Colleges and Universities, 9*(3), 105–109. (in Chinese).

Yang, Y., Zhao, Q. (2013). On master of education (MEd) training problem. *Educational Theory and Practice,* 33(6), 9–11. (in Chinese).

CHAPTER 13

SECONDARY SCIENCE TEACHER EDUCATION OF SOUTH KOREA

Jongseok Park
Kyungpook National University

Kew-Cheol Shim
Kongju National University

NATIONAL REQUIREMENTS FOR SECONDARY SCIENCE TEACHERS OF SOUTH KOREA

The training of pre-service secondary school science teachers in South Korea functions within universities' colleges of education, courses for teaching professions, and the graduate schools of education. These teacher education systems are the same in other subjects. Current pre-service secondary science teacher training institutions in South Korea are shown in Table 13.1.

Pre-service secondary science teachers are trained at 73 total universities, which provide teacher education programs if that training is not offered in the college of education, and at 50 graduate education schools in various universities. From these institutions, we discharge more than 2,000 science teachers every year.

Model Science Teacher Preparation Programs, pages 271–285

TABLE 13.1 Number of the Teacher Training Institutions of South Korea

Undergraduate		
College of Education	Course for Teaching Profession	Graduate School of Education
19	54	50

Source: Ministry of Education, Korea(2013)

Generally, secondary science teachers are trained at the college of education in a university. If university students are not trained at the college of education, but want to become a secondary science teacher, they can apply for the undergraduate course for teaching profession. The selection of pre-service students for the undergraduate course for teaching profession is made among those who have completed the second year. Only the pre-service students with the permission of the Ministry of Education can be selected based on an entrance quota. The personnel selection for the entrance quota of 10% is determined through a selection process and standards including course grades, an interview, and the test of aptitude for the teaching profession and personality.

Graduate education schools have refresher courses and training courses for secondary science teachers. It is difficult to determine the exact number of secondary science teachers who participate in them because the number of students admitted each year varies. Pre-service science teachers of training courses in the graduate school of education can acquire the certificate of secondary science teacher by completing credits related to the same subject as a bachelor's degree. Table 13.2 displays this information in accordance with pre-qualified candidates who are able to acquire qualifications of a secondary science teacher from a graduate education school.

In South Korea, secondary school science teachers were trained under the combined system of the purpose-oriented system and the open system, even though primary school teachers were fostered under the purpose-oriented system (Kim, Chung, & Kim, 2012). Primary school teacher training institutions have been operated under the purpose-oriented system, which are at independent schools (called the National University of Education) that areseparated from general universities. Secondary school teacher training institutions have been operated under both the purpose-oriented system at the College of Education at National Universities, and the Department of Education of a university, and the open system through the teaching training course in an undergraduate program and at graduate schools of education. The system of secondary science teacher training in Korea faces a problem of imbalance of supply and demand due to an oversupply of science teachers. The number of students per teacher until 1989

TABLE 13.2 Acquisition of Certificate After Graduation According to the Qualification Before the Admission

	Person Admitted Into the Graduate Education School	Certificate Acquired After Graduation
Training Course	a. Person who has graduated from the related department	A master's degree and the qualification of science teacher.
	b. Present teachers of secondary school	A master's degree and the qualification of minor science teacher.
	c. Person who doesn't meet either a or b	A master's degree.
Refresher Course	d. Present teachers of secondary school	A master's degree and the qualification of minor science teacher.
	e. Person who doesn't meet d of the above	A master's degree.

was more than 40 in middle schools and more than 30 in high schools (MOE and KEDI, 1997). In the early 2000s, the number of high school students per teacher was reduced by fewer than 20 (18.32 students); for 2010s, the number of students is around 17. Thus, as the school-age population decreases the number of appointed secondary pre-science teachers has decreased. The training system of secondary school science teachers faces a lack of a framework for improving the quality of science teachers. Pre-service secondary science teachers have been selected in the same way as students of subjects from other colleges of university have been. There are some criticisms of this system such as too much formal interview, and the test of aptitude and personal test for selecting pre-service science teachers of the college of education, and the graduate school of education.

ACCREDITATION OR NATIONAL ASSESSMENTS REQUIRED OF SOUTH KOREA FOR SCIENCE TEACHER PREPARATION

Secondary school science teacher certification is divided into 5 certificates of common sciences: physics, chemistry, biology, earth science, and common science. Pre-service secondary science teachers are granted the second-grade certificate of secondary science teacher without national assessments if they complete national qualifications of secondary science teachers at the college of education, the undergraduate course for teaching profession, and the graduate education school. We have two kinds of secondary science teacher certificates; the second-grade and the first-grade. Pre-service secondary science teachers can acquire the second-grade or the first-grade certificate after meeting the qualifying criteria shown in the Table 13.3. The first-grade certificate can be acquired in secondary school for

TABLE 13.3 Criteria of Secondary Science Teacher for the Second Grade and the First Grade Qualifications

	Second-Grade Science Teacher	First-Grade Science Teacher
Criteria of Qualifications	1. People who graduated the college of Education. 2. People who got Master's degree at Graduate school. or Education department of the graduate school selected by the Minister of Education. 3. People who completed the temporary teacher training course. 4. People who graduated Education department. 5. People who got certain credits for teaching profession courses as an graduate of university, college of industry 6. People who have an assistance teacher license of secondary-school with work experiences more than 2 years and has reeducated. (Assistant teacher license has been abolished since class of '82) 7. People who graduated university and has an assistant teacher license of primary school. 8. An assistant professor at college of education or junior college with more than two years of teaching experiences. 9. Possessing qualifications of industry-educational adjunct teachers in accordance with the provision of Article 22(Except honorary professors), people who got credits of the teaching profession subjects selected by a presidential decree at the university selected by the Minister of education or teacher in-service training institute by passing through the recommendation of appointment authority holder and the screening of superintendent of education. In this case, the subject person recommendation screening standard of "appointment authority holder" and screening standard of superintendent of education will be decided by the president.	1. People who has regular teacher's license with teaching experiences at least one year and got a master's degree from education department selected by graduate school of education or the Minister of Education. 2. People who don't have regular teacher's license but got secondary school teacher's license(secondary) from the Minister of Education and people who have teaching experiences at least 3 years after getting master's degree from graduate school of education department selected by graduate school of education or the Minister of Education. 3. People who have secondary school regular teacher's license with teaching experiences at least 3 years and who received certain reeducation. 4. Professors and associate professors of teacher's college and junior college with teaching experiences(at least 3 years). ☐ *The graduate school selected by the Minister of education* (the enforcement regulations on certification of teachers, the provisions of Article 16) 1. Graduate school of Seoul National University 2. Graduate school of Korea National University of Education 3. Graduate school of teacher's college admitted and installed as professional graduate school of education

three years after winning the second-grade science teacher certificate and completing the first-grade training course.

The Ministry of Education in South Korea awards the secondary science teacher certificate without national assessment to persons who have completed the qualification criteria. The qualification criteria necessary to acquire the secondary science teacher certificate without national assessment consists of the liberal arts course, the major course, the general pedagogical course, and the Teaching Aptitude and Personality Test (Table 13.4). The major course includes the basic content knowledge course, the pedagogical content knowledge course, and the general pedagogical course which includes the education theory course and the field experience course.

In general, subjects of liberal arts course for pre-service secondary science teachers are to be completed in the first grade, and general pedagogical subjects and majors have to be completed in the 2nd to 4th grade. The qualification criteria without national assessment to acquire the secondary science teacher certificate at the graduate education school are the same as at the college of education and teaching course of the university (Table 13.5)

TABLE 13.4 Qualification Criteria of Secondary Science Teachers Certificate Without National Assessment at the College of Education and the Course for Teaching Profession

	Incoming Students of 2009–2012	Incoming Students From 2013
Major (Science)	• More than 50 credits – More than 21 credits (7 subjects) of basic content (science) – More than 8 credits (3 subjects) of pedagogical content education area	• More than 50 credits – More than 21 credits (7 subjects) of basic content (science) – More than 8 credits (3 subjects) of pedagogical content education area
General Pedagogy	• More than 22 credits – More than 14 credits (7 subjects) of the pedagogical theory – More than 4 credits (2 subjects) of the pedagogical literacy – More than 4 credits of field experience (including 2 credits of educational service activities)	• More than 22 credits – More than 12 credits (6 subjects) of the pedagogical theory – More than 6 credits (3 subjects) of the pedagogical literacy – More than 4 credits of field experience (including 2 credits of educational service activities)
Standard of Grading	• The average of the total graduation grade : more than 75/100	• The average grade of the major subjects: more than 75/100 • The average grade of Teaching profession subjects (including practice teaching): more than 80/100
Et Cetera	• Getting eligibility judgement at the test of teaching aptitude and personality checking at least once	• Getting eligibility judgement at the test of teaching aptitude and personality checking at least twice

but are different from credits of basic content knowledge course and the pedagogical content knowledge course. Such a difference is because of the courses taken at the undergraduate level.

The curriculum for pre-service secondary science teachers lacks adhesion to secondary school and has a lack of practical knowledge that can be used directly in the individual schools (Kim, Park, & Mun, 2009; Yang, Kwak, Han, & Noh, 2013). The curriculum of science teacher training institutions is designed with a focus on pedagogical theory and science contents. Teacher training courses in developing and operating a wide range of subjects associated with the field are required. It is necessary to develop a teacher training curriculum of teacher training institutions closely linked to the secondary school fields (Kim & Lee, 2006; Shim, 2010). In South Korea, the government can entrust autonomous efforts of secondary teacher training institutions to ensure the quality of secondary science teachers. It is necessary to devise a plan to thoroughly assess the quality of teacher training.

TABLE 13.5 Qualification Criteria of Secondary Science Teachers Certificate Without National Assessment at the Graduate Education School

	Incoming Students of 2009–2012	Incoming Students From 2013
Major (Science)	• More than 50 credits – More than 14 credits (5 subjects) of basic content (science) – More than 6 credits (2 subjects) of pedagogical content education area	• More than 50 credits – More than 14 credits (5 subjects) of basic content (science) – More than 6 credits (2 subjects) of pedagogical content education area
General Pedagogy	• More than 22 credits – More than 14 credits (7 subjects) of the pedagogical theory – More than 4 credits (2 subjects) of the pedagogical literacy – More than 4 credits of field experience (including 2 credits of educational service activities)	• More than 22 credits – More than 12 credits (6 subjects) of the pedagogical theory – More than 6 credits (3 subjects) of the pedagogical literacy – More than 4 credits of field experience (including 2 credits of educational service activities)
Standard of Grading	• The average of the total graduation grade: more than 75/100	• The average grade of the major subjects: more than 75/100 • The average grade of Teaching profession subjects (including practice teaching): more than 80/100 ☐ Only the grades while attending at the college of education will be calculated
Et Cetera	• Getting eligibility judgement at the test of teaching aptitude and personality checking at least once	• Getting eligibility judgement at the test of teaching aptitude and personality checking at least twice

APPOINTMENT AS A SECONDARY SCIENCE TEACHER

The appointment of secondary teachers including secondary science teachers depends on government-established secondary schools including national and provincial schools, and nongovernment-established schools, even though they are open to the public. New science teachers are appointed through the open competitive examination called the Examination for Appointing Secondary School Science Teachers (EASST).

For government-established schools, successful applicants are appointed to government-established secondary schools as a secondary science teacher through the EASST, which is organized by the education offices of metropolitan cities and provinces. For non-government-established schools, new secondary science teachers are appointed through the EASST, which can be developed and organized to recruit excellent ones by the committee of each individual school.

The EASST for government-established schools consists of two phases. The first phase of the EASST includes the examination on the major area for the basic content knowledge and the pedagogical content knowledge, and the general pedagogical area. It has been criticized that the examination was based on knowledge centered questions which cannot reflect the holistic form of teacher competency. The government has tried to enhance the teaching competency in the second phase of the EASST. The second phase includes the examination on the ability of creating lesson plans relevant to teaching-learning scenarios, teaching practices, experiments, and interviews. The second phase can be carried out independently on provincial education offices (Table 13.6). For example, the education office of Gyeonggi province introduced group discussion interviews about lessons through discussions about connections to teaching demonstrations. The Kangwon provincial education office asks candidates to do teaching demonstrations without creating teaching-learning scenarios and introduced oral statement style interviews requiring a prompt reply provided from the Kangwon education office. Daegu metropolitan city will establish the literacy of liberal arts and mental attainment as new items of interview evaluation in 2016 : the Analects of Confucius, Emile, and Mingxin Baojian.

On the other hand, appointing teachers through the EASST to screen a limited personnel for placement in public schools might contribute to improving the quality of education (Shin, 2002). It was thought that efforts to select competent teachers would positively affect the academic atmosphere in the teacher training institutions. However, the rate of competition with regards to the teacher appointment exam is very high, and this creates other unexpected side-effects. For example, there is a larger number of certificate holders than a demand for secondary science teachers, and they can become a teacher through competitive examination (MOEST, 2010). The government

TABLE 13.6 The Second Phase of the EASST

Education Office	Experiment	Creating Lesson Plans	Teaching Demonstration	Interview of Teaching Aptitude
Seoul	O	O	O	O
Busan	×	O	O	O
Daegu	O	O	O	O
Incheon	×	O	O	O
Gwangju	×	×	O	O
Daejeon	O	O	O	O
Ulsan		O	O	O
Gyeonggi	×	×	O	O
Kangwon	×	O	O	O
Chungbuk	×	O	O	O
Chungnam	×	O	O	O
Jeonbuk	×	O	O	O
Jeonnam	×	O	O	O
Gyeongbuk	O	O	O	O
Gyeongnam	×	O	O	O
Jeju	×	O	O	O
Sejong	×	O	O	O

has spawned efforts to resolve the problem of overabundance of pre-service secondary science teachers by reducing the number of students for the certification through evaluating the teacher training institutions. This has enabled programs to improve the quality of their teacher training systems. Finally, these reforms can help enhance the quality of the teacher training systems at the national level and initiate overall restructuring (Gu et al., 2009).

CONTENT REQUIREMENTS FOR DISCIPLINE LICENSE

Pre-service secondary science teachers have to complete basic content subject courses for the qualifications of secondary school science teachers, which depend on five majors of common sciences, physics, chemistry, biology, and earth science (Table 13.7). Pre-service secondary science teachers should complete these courses as designated by the pursued certificate. Pre-service secondary science teachers who are majoring one of physics, chemistry, biology, and earth science can get the second-major license of common science. They complete more than 50 credits of their main major, then complete credits within the required courses.

TABLE 13.7 Basic Content Courses for the Qualification of Secondary Teachers of Common Sciences, Physics, Chemistry, Biology, and Earth Science

Subjects	Related Departments	Required Courses
Common Science	Science Education, Physical Education, Chemistry Education, Biology Education, Earth Science Education, Common Science Education Department, and the related departments	1. common science education(or science education) 2. completing at least 2 subjects of each field of a–d (except subjects of main major field). a. general physics and experiment, electromagnetism, and modern physics b. general chemistry and experiment, inorganic chemistry, and organic chemistry c. general biology and experiment, cell biology, and molecular biology d. general earth science and experiment, geology, and atmosphere science
Physics	Science Education, Physical Education, Physical Department, and the related departments	dynamics, quantum mechanics, electromagnetism, heat and statistics, physics and optical, electronic wave physics, physics, and physical education experiment
Chemistry	Science Education, Chemistry Education, Chemistry Department, and the related departments	physical chemistry, physical chemistry experiment, organic chemistry, organic chemistry experiment, inorganic chemistry, inorganic chemistry experiment, analytical chemistry, and analytical chemistry experiment
Biology	Science Education, Biology Education, Biology Department, and the related departments	Cell biology, embryology plant, physiology, animal physiology, genetics, taxonomy, ecology, molecular biology, microbiology and biochemistry
Earth Science	Science Education, Earth Science Education, Earth Science Department, and the related departments	geology, astronomy, atmospheric science, oceanography, geophysics, earth and environmental science, and natural disasters and energy resources

But, after pre-service secondary science teachers who are majoring only one of physics, chemistry, biology, and earth science are appointed at schools, they should teach middle school science and high school general science, as well as subjects of their main major field. Yet, a lot of science teachers tend to avoid teaching common science as they prefer to teach subjects in their main major fields (Yang et al., 2013). That is, they feel pressured to teach both middle school science and high school general science (Shim & Song, 2015; Song, Hong, Kim, Han, & Shim, 2012). Because they are required to prepare for the main major of the examination for appointing secondary school science teachers, they have a tendency to avoid selecting second-major license of common science.

In addition, there has been a request to improve teacher training courses for common science teachers. Most of common science teacher training courses for the pre-service secondary science teachers have been operated simply as a form of combined four science fields including physics, chemistry, biology, and earth sciences at teacher training institutions (Son, 2009). Science teacher training courses should be operated to get competencies of teaching general sciences and also to understand appropriate science contents for middle and high school sciences.

GENERAL PEDAGOGICAL COURSES REQUIRED FOR SCIENCE TEACHER PREPARATION OF SOUTH KOREA

The general pedagogical course required for secondary science teachers should have at least 22 credits including at least 12 credits of pedagogical theory course, at least 6 credits of pedagogical literacy course, and at least 4 credits of field experience course (Table 13.8). The pedagogical theory course has subjects such as introduction to education, philosophy and history of education, curriculum, educational evaluation, sociology of education, education psychology, educational administration and management, life coaching and counseling, and other subjects on teaching theory. The pedagogical literacy course has subjects such as introduction to special education including content related to gifted learners, teacher's work practices, and theory and practice of school violence prevention. The field experience course has practical subjects such as internship for teaching practice and educational service.

THE PEDAGOGICAL CONTENT COURSES REQUIRED FOR SCIENCE TEACHER PREPARATION

Pre-service science teachers complete at least 6 credits in the pedagogical content education area including at least 3 credits of science education theory as major required courses for qualification without national assessment. Generally, pre-science secondary science teacher training institutions have opened the lectures of science education theory as well as science teaching and learning method, science logic and writing, and research for science learning materials. Before 2009, the pedagogical content course and the course for teaching profession were not separated in the area of the EASST. Since 2009, the pedagogical content course has been separated with the course for teaching profession and included in major subjects that should be mandatory for at least 8 credits.

Pre-service science teachers learn science curriculum, history and philosophy of science, science inquiry, science education theory, science teaching

TABLE 13.8 General Pedagogical Course Required for Secondary Science Teachers

	Incoming Students of 2009–2012	Incoming Students Since 2013
Pedagogical Theory	• More than 14 credits (more than 7 subjects) – General Education Outline – Educational Philosophy and Educational History – Course of Education – Assessment of Education – Teaching Methods and Educational Technology – Educational Psychology – Educational Society – Educational Administration and Educational Management – Guidance of Life and Counseling – Other Subjects About Teaching Profession Theory	• More than 12 credits (more than 6 subjects) – General Education Outline – Educational Philosophy and Educational History – Course of Education – Assessment of Education – Teaching Methods and Educational Technology – Educational Psychology – Educational Society – Educational Administration and Educational Management – Guidance of Life and Counseling – Other Subjects About Teaching Profession Theory
Pedagogical Literacy	• More than 4 credits (more than 2 subjects) – Introduction to Special Education (more than 2 credits) □ Including area of Special Education for the Gifted – Practice of Teacher's Work (more than 2 credits)	• More than 6 credits (more than 3 subjects) – Introduction to Special Education(more than 2 credits) □ Including area of Special Education for the Gifted – Practice of Teacher's Work (more than 2 credits) – The Theory and Practice of School Violence Prevention (more than 2 credits)
Field Experience	• More than 4 credits (more than 2 subjects) – Internship for Teaching Practice (more than 2 credits) – Educational Service (more than 2 credits)	• More than 4 credits (more than 2 subjects) – Internship for teaching practice (more than 2 credits) – Educational service (more than 2 credits)

and learning, and evaluation and assessment of science learning in preparation of the EASST through a pedagogical content course (Table 13.9). Science education theory and science logic and writing are mandatory for qualification of secondary science teachers.

The pedagogical content course comprises a very low proportion of credits required for the qualification of secondary science teachers. Most pedagogical content included in the subject of science education theory have been taught for pre-service science teachers. Science education experts point out that there are needs for opening subjects such as science inquiry and scientific thinking, science curriculum, science teaching and

TABLE 13.9 Pedagogical Contents of the Examination for Appointing Secondary School Science Teachers

Area	Short-Range
Science Curriculum	Characteristic/purpose/teaching-learning method of science education curriculum, transition of science education curriculum, content element of education curriculum, education curriculum of foreign countries, etc.
History and Philosophy of Science	Type and formation of scientific knowledge (fact, concept, principle, rule and theory etc.), nature of science, philosophy of science (positivism, rationalism, empiricism, Popper's falsification, Kuhn's scientific revolution, Lakatos's research program etc.), history of science, etc.
Science Inquiry	Science inquiry process (basic/integrated process), scientific thinking (creative thinking, logical thinking. theory-laden observation, analogy, meta-cognition), scientific reasoning method (deduction, induction, hypothetical deductive method, abduction), how to use experimental equipment and tools, laboratory safety, etc.
Science Education Theory	Constructionism, Piaget's theory of cognitive development, Ausubel's meaningful learning theory, Bruner's instructional theory, Vygotskian theories, the conceptual change theory (Driver, Hashweh, Posner, West and Pines etc.), student's misconception, understanding of scientific concept in the context of class, etc.
Teaching and Learning of Science	STS teaching-learning method, cooperative learning, V map, concept map, questioning method, discussion method, scientific treatment (writing), teaching and learning model (Karplus's learning cycle model, POE, PEOE, 5E instructional model, Lawson's learning cycle model, generative learning model , discovery learning model etc.), etc.
Evaluation and Assessment of Science Learning	Klopfer's Taxonomy of Education Objectives, knowledge of science, scientific inquiry ability, interest in science, attitude related to science, content validity/credibility/discrimination power, etc.

Source: Lee et al., 2013

learning theory and practice, science teaching and learning materials development, science learning assessment, and science education facilities and environments (Kim et al., 2009).

SPECIFIC FIELD EXPERIENCE REQUIREMENTS

Pre-service secondary science teachers should complete field experience courses including at least 2 credits of internship for teaching practice, and at least 2 credits of educational service. Field experience courses belong to the general pedagogical area. Through an internship for teaching practice, pre-service secondary science teachers observe teaching practices

TABLE 13.10 Subjects and Contents of the Field Experience Course for Pre-Service Secondary Science Teachers

Subject	Credit	Contents
School Internship Program	More Than 2	• Time reference per credit – Carried out as a full-time system → 2 week per 1 credit (it doesn't matter if holidays are included in the term) (If the period of actual training is not in succession, regard 80 hours per 1credit as standard) • The period of practice teaching – University will make an autonomous decision on the period of implement → first semester of senior year
Educational Service Activities	More Than 2	• The meaning of educational service activities – Serving the talent of university students to kindergartener, elementary, junior high, and high school students by educational methods(teaching assistant, underachiever leading, after school instructors) • Time reference per credit – Regard more than 30 hours per hour as standard • Institute which is possible of Educational service – All institutes that is available to school internship program – Social education institutions or facility which is related with major field – Non-profit organization admitted by director of public institution

and teach students at the secondary school for 4–6 weeks (Table 13.10). They earn credit of educational service through performing educational service activities through a variety of roles such as secondary teacher, student guidance of underachiever, and after-school teacher. The internship for teaching practice is to apply the theories and techniques taught in the curriculum of teacher training courses and to practice at the last step of pre-service teacher education. The process of managing internships for teaching practice consists of pre-guidance for internship, pre-meeting with the tutor, practice school visit meeting with student teachers and the tutor, consultation of teaching practice, an interview about internship program, and surveying about the satisfaction of trainee (Table 13.11).

Pre-service secondary science teachers can apply what they learned at the college of education through the internship program at school. In addition, they should continue to polish and expand the qualities needed to become secondary science teachers through internship. But, the period of internship for teaching practice at school is not sufficient. The internship period at school needs to be extended in order to improve educational expertise.

TABLE 13.11 Process of Internship Program at School

Before Internship	In the Middle of Internship	After Internship
• Pre-guidance for internship • Pre-meeting with the tutor about internship program • Advance guidance to student teacher	• Visit meeting with student teachers and tutor • Evaluation group meeting and consulting after observing the teaching practice	• Interview about internship program • Surveying about the satisfaction of trainee

SUGGESTION FOR IMPROVEMENTS OF SECONDARY SCIENCE TEACHER PREPARATION

Pre-service secondary science teachers were at the top of their class as high school students. They were within the top 30% of academic achievement in the Korean SAT when they were in high school. Compared to the commitment of teacher resources and talents, the lack of ability to foster the competent teachers has been being criticized (Kim et al., 2012). This suggests that the framework of secondary science teacher training courses, and the training system for secondary science teacher should be improved.

The portion of content knowledge course of science teacher training curriculum is higher than that of pedagogical content course. Simply, not to increase the proportion of pedagogical content course, the discussion about practical subjects related to the actual learning required in schools, and a course to improve competencies related to instruction and to develop skills of prospective teachers is required (Chung et al., 2010). Pre-service secondary science teachers should recognize the gap between theory and reality at schools through the internship program, and a plan to reduce the gap is needed. For this, the plan may be made in conjunction with the professor through active communication with the teachers mentoring student teachers at school.

In conclusion, science teacher training courses at the college of education will be improved by strengthening the teaching practical skills (Kim et al., 2012; Park, 2007). Elementary and secondary school teachers via separate teacher training institutions were trained for national qualification. It is necessary to train both of them in the same teacher training institution. If elementary and secondary science teachers are trained through an integrated operational and training curriculum in the same institutions, training can provide even greater academic knowledge and understanding about students.

REFERENCES

Chung, M. K., Kim, K. S., Ryoo, J. S., Kim, B. C., Park, S. W., & Moon, C. S. (2010). *Improving teacher education curriculum*. Seoul, Republic of Korea: Korea Education Development Institute.

Gu, J., Kim, K., Kim, W., Kim, K., Kim, J., Namgoong, J., . . . & Hong, C. (2009). *The study for strengthening the third phase evaluation of teacher training institutes*. Seoul, Republic of Korea: Korea Education Development Institute.

Kim, G. S., Chung, M. K., & Kim, D. K. (2012). *Successful strategy for training teachers in Korean education*. Seoul, Republic of Korea: The Ministry of Strategy and Finance.

Kim, Y., Park, J. S., & Mun, J. (2009). Reconsideration on current curriculum for science teacher preparation based-on standards for professional science teacher. *School Science Journal, 3*(1), 48–58.

Kim, J. H., & Lee, K. Y. (2006). Investigation of the earth science teacher education programs in the college of education and their improvement plans. *Journal of the Korean Earth Science Society, 27*(4), 390–400.

Lee, B., Shim, K. C., Shin, M. K., Kim, J., Choi, J., Park, E., . . . & Kim, Y. J. (2013). Analyses of science education theories in the question items of the examination for appointing secondary school science teachers. *Journal of Korean Association for Science Education, 33*(4), 794–806.

Ministry of Education, Korea. (2013). *Current status of 2013 teacher training institution*. Seoul, Republic of Korea: The Ministry of Education.

Ministry of Education, Korea & Korea Educational Development Institute. (1997). *Traces of Korean education in statistics*. Seoul, Republic of Korea: Korea Educational Development Institute.

Ministry of Education, Science, and Technology, Korea. (2010). *2010 working guide to teachers license examination*. Seoul, Republic of Korea: Ministry of Education, Science and Technology.

Park, S. W. (2007). A study on the curriculum of teacher education programs: Issues and possible directions. *Journal of Korean Teacher Education, 24*(2), 143–173.

Shin, G. (2002). Research in effectiveness analysis of new appointment policy of primary school teachers. *Education Research, 40*(4), 219–241.

Shim, K. C. (2010). Study on initial teacher training programmes for science teachers in England: Roehampton University, England. *Biology Education, 38*(3), 492–506.

Shim, K. C., & Song, S. C. (2015). A study on the perception of science teachers about convergence "science" for high school students by experiences of teaching and professional development. *Biology Education, 43*(2), 158–169.

Son, Y. A. (2009). The understanding and reality of operation of science integrated curriculum. *The Program of Conference of the Korean Society for the Study of Curriculum Integration, 4*, 43–77.

Song, S. C., Hong B., Kim N. H., Han H. J., & Shim, K. C. (2012). Study on perceptions of high school students and science teachers about high school fusing science. *Journal of Science Education, 36*(1), 130–138.

Yang, C., Kwak, Y., Han, J., & Noh, T. (2013). Current status of teacher education curriculum and recruitment of general science teachers and ways to improve them as suggested by professors from the department of science education. *Journal of Korean Association for Science Education, 33*(2), 345–358.

CHAPTER 14

SECONDARY SCIENCE TEACHER EDUCATION/ TRAINING IN JAPAN

Tetsuo Isozaki
Hiroshima University

Takuya Ochi
Hiroshima University (PhD course)

This chapter explores the provision of secondary science teacher education/training in Japan. It begins with a historical review of this provision followed by a description of the current teacher education/training situation. It will then critically discuss the implications and future of how science teachers are prepared in Japan to teach. In Japan, the central government, partially the Ministry of Education, Culture, Sports, Science, and Technology (MEXT: *Monbukagakushō,* in Japanese) has a powerful influence in the field of education, as well as other fields. Prior to 2001, the MEXT was known as the Ministry of Education, Science, and Culture. However, the current term MEXT will be used in this chapter.

There are 47 prefectures (provinces) in Japan, and 20 cities designated by government ordinance. Prefectures and some designated cities have

Model Science Teacher Preparation Programs, pages 287–306

their own board of education. Among the MEXT's major roles in teacher education/training are making national policies and regulations on education, and designating teacher education/training institutions. The board of education, on the other hand, is executive in function. It manages the affairs of education at the local level and also grants a teaching certificate to those who meet the certificate requirements.

There is a distinct difference in meaning between the terms *teacher education* and *teacher training* (Peters, 1966; Aldrich, 1996). Both terms are found in some MEXT documents. Although we have to define the meaning of the two terms and describe the difference between them, it is very hard for us to discuss about it briefly. In this chapter, however, we use the term "training" to mean "education," but use the term "education" to refer to both pre- and in-service teacher education/training.

BRIEF HISTORY AND CURRENT SYSTEM OF SECONDARY SCIENCE TEACHER TRAINING IN JAPAN

Teacher training in Japan started with the establishment of normal schools in 1872. The normal school was established on the principles of modern education in conformity with the Education System Order (Ministry of Education, Science and Culture, 1980). The higher normal schools that educated secondary school teachers (boys' secondary schools, girls' secondary schools and normal schools for elementary teachers) were established in Tokyo in 1886 and in Hiroshima in 1902. Elementary school teachers were trained at normal schools in every prefecture. There were three ways to be a secondary school science teacher: graduating from a higher normal school; graduating from a university designated by the Ministry of Education as an institution of teacher training; and passing the examination regulated by the Ministry of Education. Before World War II, there was a controversy around secondary school teachers in relation to "professionalism" and "academism" (Isozaki, 2001). A "professional" teacher was categorized as someone who graduated from a higher normal school and learned both pedagogical knowledge and content knowledge. In contrast, an "academic" teacher was a university graduate, especially from an imperial university, who had learned only deep content knowledge without pedagogical knowledge. "Professional" teachers were criticized for their lack of deep content knowledge, while the "academic" teachers were criticized for their lack of pedagogical knowledge. The two types of teachers were sometimes antagonistic towards each other. For the elementary school teachers, they were trained only in prefectural normal schools, which were considered a "closed system." Teachers who graduated from normal schools and partially graduated from higher normal schools were derided as "normal school types" and were criticized for "hypocrisy, formalism, unadaptability, and indecision" (Kumura and Iwahashi, 1967, p. 82).

The current teacher training system was inaugurated in 1949 based on recommendations of the U.S. Education Mission during the Occupation period (1945 to 1952). As a result, the traditional normal and higher normal schools were abolished. There are now two principles for teacher training after World War II: "teacher training at universities" and "open system" for teachers' certificates as a "teaching license." The university-based teacher training implies that the basic requirement for school teachers teaching from kindergarten to upper secondary school is a four-year bachelor degree. The purpose for this minimum requirement is to produce diverse human resources with highly specialized content and pedagogical knowledge and skills. The second principle means that if a national, municipal, or private university has been accredited as a teacher training institution by the MEXT and has been treated and regarded as equal by the enactment of the Educational Personnel Certification Act, it could provide various kinds of teaching certificates. The curriculum for pre-service teacher training in these institutions should be developed within the framework of this Act and other regulations by the MEXT.

Pre-service teacher training and its curriculum in Japan have responded to changes in society after World War II. These changes are evident in the balance between the supply and demand of teachers, children's social environment, increased demand for life-long learning, globalization efforts, and so on.

This brief history of teacher training in Japan shows that the present teacher training takes the form of a unified pattern based on two principles that were provided after World War II. Its characteristics can be summarized as follows:

1. The MEXT's responsibilities encompass teacher education as well as all education fields. There is *no* national curriculum/set of standards for pre-service teacher training, however, all teacher training institutions should be accredited and designated based on the laws and regulations of the MEXT (see Figure 14.1).
2. In Japan, *all* full-time and part-time teachers from kindergarten to upper secondary school (K to G12) must have teaching certificates. There are three teaching certificates which differ in acquisition method and validity: the regular teaching certificate; the special teaching certificate; and the temporary teaching certificate. The regular teaching certificate is the most popular and is issued by every prefectural board of education. It is valid for all prefectures of Japan. However, it is only valid for 10 years. Teachers must renew their teaching certificates by taking a course offered by universities

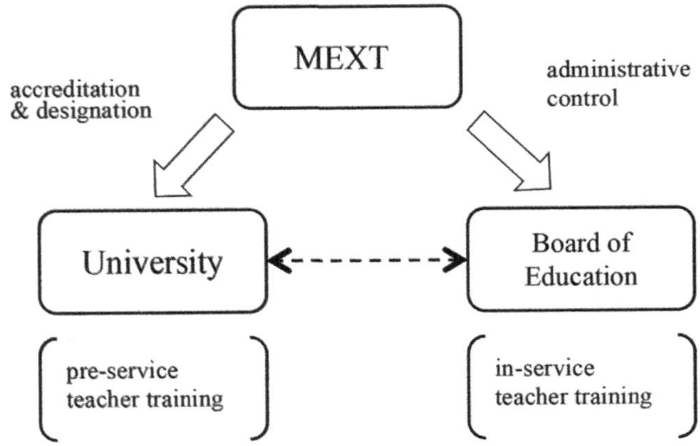

Figure 14.1 Teacher education framework in Japan.

and other institutions designated by the MEXT. The regular teaching certificates are categorized as specialized, class I, and class II based on teachers' academic background. A four-year bachelor's degree is the basic qualification for the class I certificate and a two-year master's degree for specialized certificates. Secondary school teachers' certificates are valid for *one* subject such as science (not separated into physics, chemistry, biology and earth science) and mathematics.

3. Under the "open system," teaching certificates are granted to anyone who completes the required courses in a university. However, this system is considered a double-edged sword. Compared to the system before World War II, the current system is more advantageous in terms of helping Japan meet the demand for teachers and produce diverse human resources in the field of education. Regrettably, the "open system" has caused a definite lowering of certification standard. The actual organization of the pre-service teacher training curriculum differs depending on the characteristics of institutions. As Table 14.1 indicates, over 80% of teacher training institutions offering lower and upper secondary science certificates are non-education faculties and departments. There are many students that do not belong to education faculties, who are only interested in getting teaching certificates. Their intention is to take the minimum number of credits required for a teaching qualification in order to enter the labor market. Therefore, it can be ascertained that economic conditions influence the number of prospective science teachers. Nevertheless, they still have to undergo teaching practice, which is compulsory at

TABLE 14.1 The Number of Institutions for Secondary School Science Teacher Certificates

	Lower Secondary Science School Certificate				Upper Secondary Science School Certificate			
	Number of Universities	Number of Faculties	Educational Course	Scientific Course	Number of Universities	Number of Faculties	Educational Course	Scientific Course
National University	64	124	46 (37%)	78 (63%)	67	158	45 (28%)	113 (72%)
Municipal University	13	16	0 (0%)	16 (100%)	16	22	0 (0%)	22 (100%)
Private University	88	145	8 (6%)	137 (94%)	92	152	7 (5%)	145 (95%)
Total	165	285	54 (19%)	231 (81%)	175	332	52 (16%)	280 (84%)

Notes: The total number of universities in Japan: National (86), Municipal (89), Private (604). Education course means faculty of education or department of science education, scientific course means scientific faculties and departments.

Source: MEXT (2015). *Universities that provide teacher certificates in 2015*. Retrieved from http://www.mext.go.jp/component/a_menu/education/detail/__icsFiles/afieldfile/2015/12/01/1287058_1.pdf

MEXT (2016). *School Statistics in 2016*. Retrieved from http://www.mext.go.jp/b_menu/toukei/002/002b/1368900.htm.

Calculated by the authors.

their alma maters. This is criticized and ridiculed by teachers and media as "teaching practice pollution" (Iwata, 2006).

The brief history and highlights of the current system provided in this chapter elucidate the present teacher training system and the multi-faceted relationships between schools, government policies, and universities in a climate of socio-political changes.

TEACHING PROFESSION

School System and Curriculum

One of the distinctive features of the education system in Japan is that schools offer a common curriculum which is general education-oriented for all students under a single-track system. After World War II, based on recommendations of the U.S. Education Commission, a 6-3-3-4 school system was introduced. Elementary school is 6 years (G1 to G6: 6–11 years old) and lower secondary or junior high school is 3 years (G7 to G9: 12–15 years old). Both are compulsory education and offer general education.

The elementary school curriculum consists of compulsory subjects such as Japanese language, social studies, and science known as *rika* in Japanese (Ogawa, 2015). Moral education, the period of integrating studies, special activities, and foreign language activities are also included. Lower secondary school curriculum also consists of compulsory subjects such as Japanese language, social studies, science, foreign language, moral education, the period of integrating studies, and special activities. Upper secondary or senior high school (G10 to G12: 16–18 years old) is not compulsory but enrollment rate is more than about 98% in 2012 (MEXT, 2013). At upper secondary education level, general, vocational and specialized programs (agriculture, home economics, and other fields of study) are provided. The general program at upper secondary school curriculum consists of compulsory and elective subjects such as Japanese language, mathematics, and science. The period of integrating studies and special activities are also offered. In this level, the subject science consists of two types: balanced science and separate science. The first type is "science and human lives," while the second one consists of basic and advanced physics, chemistry, biology, and earth science. There is a different type of subject known as "scientific investigation activities" focused on scientific investigation. There are ten science subjects, and *all* students, even though they are not scientific students who do not choose the science track must take two or three from 10 subjects as compulsory subjects. Because Japanese elementary and secondary education are general education oriented, *every* student must take some compulsory sub-subjects (e.g., physics, world history) from *all* subject areas (Japanese language, mathematics, science, geography and history, civics, foreign language, home economics, arts, and health and physical education). Of course, scientific students should take compulsory and also elective science subjects.

In Japan the course of study, like the national curriculum or standards, is implemented from elementary school to upper secondary school in every subject. Furthermore, all textbooks which are used in schools should be authorized by the MEXT along with the course of study.

Multiple-Task Work of Teachers

There is no doubt that the most important work of teachers in schools is to teach their subject effectively. However, the 2013 Teaching and Learning International Survey (TALIS 2013) by the Organization for Economic Co-operation and Development (OECD) reveals that Japanese lower secondary school teachers report spending an average of 58 hours per week working. This statistic is a stark contrast to the working hours among the participating countries of TALIS 2013. The teachers from these countries work approximately 35 hours per week (OECD, n.d.). Japanese teachers' allocated time for teaching, planning, and marking are actually similar with other countries. The question then is why do Japanese teachers spend more time in school compared to teachers from other countries?

Japanese secondary teachers are required to teach not only their own specialized subjects but also the period of integrating studies, special activities, and moral education (only for lower secondary school). It is noteworthy that in secondary schools, teachers belong to their own subject department, and are also organized into grade units cutting across subject boundaries to deal with common issues and tasks pertaining to a particular grade level. They also have other responsibilities in school such as school management, research promotion, student counseling, school sports club instruction, and communication with people in the school community. Therefore, science teachers can understand students' characteristics as a whole, not only through their performance in science lessons but also through other work-related tasks and school activities such as pastoral care and tutoring classroom students (Isozaki & Isozaki, 2011, p. 37). As these facts show, essentially, Japanese teachers spend more hours performing a variety of other work-related tasks (OECD, n.d.).

Science Teaching in Schools

In lower secondary school, there is one subject science which is not separated and covers a wide range of natural sciences such as physics, chemistry, biology, and earth science including astronomy. Therefore, lower secondary science teachers have to teach a whole area of science regardless of their majors at university or post-graduate school. As mentioned previously, upper secondary school offers 10 science subjects, and there is a minimum requirement of combinations of science subjects for all students. Upper secondary school science teachers should teach some science subjects regardless of their major at university. This is because their teaching certificate is "science," not separate science subjects.

It is noteworthy that the MEXT, with the Japan Science and Technology Agency (JST), designates qualifying upper secondary schools as Super Science High Schools (SSHs). The SSHs can receive extra funding from the JST to further enhance their science and mathematics curricula. Their advanced curricula should be reflected in the school's policy and should be distinct compared to general programs. SSHs are encouraged to establish linkages with higher education institutions, and national and private research institutions for their students' scientific instruction. SSH is a national strategy for enhancing science education, and supporting the talented science education programs in Japan. At present, there is no special training program or subject provided in teacher training institutions to become a science teacher of SSHs.

TEACHER QUALIFICATION AND THE CURRICULUM FOR SECONDARY SCIENCE TEACHER TRAINING

Teacher Certificates

Table 14.2 shows the type of teaching certificates (licenses), basic requirements, and the minimum number of credits for a teaching certificate.

The total number of teaching subjects has been declining with every amendment of the law. For example, prior to 1998 the minimum number of credits for subject studies both for lower and upper secondary school teachers was 40 credits. In contrast, the minimum credits of pedagogical studies including teaching practice have been increasing from 19 credits (3 for teaching practice) to 31 credits (5 for teaching practice). Surprisingly, while the requirements of pedagogical studies for secondary schools, especially upper secondary school certifications in the current system are more than doubled before 1988, the requirements of subject studies in the current system are just half prior to 1988 (e.g., Shimahara, 2002). What do these facts mean? As Tables 14.2 and 14.3 show, there is a stronger emphasis on pedagogical studies. Among the main reasons for these changes are issues relating to student' behavior and school environment problems such as bullying and school violence. There is also a need to better support children who require special assistance and to strengthen the capacity of

TABLE 14.2 Type of Teaching Certificate and Minimum Number of Credits Since 1998

Type of Teaching Certificate		Basic Requirement	Minimum Number of Credits		
			Teaching Subjects	Pedagogical Subjects (Including Teaching Practice)	Teaching Subjects or Pedagogical Subjects
Lower Secondary School	Specialized	Master's degree	20	31	32
	Class I	Bachelor's degree	20	31	8
	Class II	2 years college degree	10	21	4
Upper Secondary School	Specialized	Master's degree	20	23	40
	Class I	Bachelor's degree	20	23	16

Source: MEXT (n.d.). Education Personnel Certification Act and regulations. http://www.mext.go.jp/a_menu/koutou/kyoin/1268593.htm)

TABLE 14.3 The Minimum Requirement of Professional and Pedagogical Studies Since 1998

Subjects	Contents of Subjects	Lower Secondary	Upper Secondary
Subjects related to significance and nature of teaching profession	• The significance and nature of teaching profession, and the role of teachers • Job specifications of the teacher • Providing opportunities of career decision	(Specialized and Class I) 2	(Specialized and Class I) 2
Subjects related to basic theories of education	• Ideas of education, history and thought of education • Mental and physical development, and learning process of children, students • Social, institutional and managerial affairs of education	(Specialized and Class I) 6	(Specialized and Class I) 6
Subjects related to curriculum and teaching methods	• The significance of curriculum and methods of its organizing • Methods of subject teaching • Methods of moral education • Methods of special activities • Methods of education and technique including ICT	(Specialized and Class I) 12	(Specialized and Class I) 6
Subjects related to student guidance, educational counseling, and career guidance	• Theories and methods of student guidance • Theories and methods of educational counseling • Theories and methods of career guidance	(Specialized and Class I) 4	(Specialized and Class I) 4
	• Teaching practice	5	3
	• Seminar in teaching profession*	2	2

* shows that this subject is introduced in 2008, and the subject was "integrated seminar" from 1989 to 1998.

Source: MEXT (n.d.). Education Personnel Certification Act and regulations. http://www.mext.go.jp/a_menu/koutou/kyoin/1268593.htm)

the family and community. All these issues have become national concerns since the 1980s. The first category of subjects shown in Table 14.3 is a new category (subjects related to significance and nature of teaching profession), and an outcome of a strong declaration of intention for the teaching profession. It includes love for teaching, pride as a teacher, and a sense of belongingness to the profession. These qualities are endorsed by the teacher education council (*Kyouiku-shokuin Yousei Shingikai,* in Japanese, 1997). As a result, science teachers as well as other subject teachers are required to have strong will and identity for the teaching profession rather than highly specialized content knowledge in science.

University students who wish to be a teacher are additionally required to take the following subjects: the constitution of Japan, physical education, foreign language communications, and information and communication technology (ICT). Additionally, university students who wish to take a teaching certificate leading to elementary school teaching and lower secondary school teaching must participate in nursing, assistant services, or communications with people with disabilities and elderly people for more than seven days at a social welfare or special needs institution.

Overview of Curriculum for Pre-Service Teacher Training and Knowledge Base

Table 14.3 shows the pre-service teacher training curriculum of pedagogical studies as required by the MEXT.

To further elaborate on the characteristics of the pre-service teacher training curriculum, Shulman's (1987) classification of the knowledge base for teaching will be used as a framework. Shulman's (1987) framework consists of seven types of knowledge: content knowledge, general pedagogical knowledge, curriculum knowledge, pedagogical content knowledge (PCK), knowledge of learners and their characteristics, knowledge of educational contexts, and knowledge of educational ends. Basically, the curriculum of pre-service teacher training courses includes *all* aspects of the knowledge base mentioned.

According to the Education Personnel Certification Act and regulations, university students who wish to be science teachers must take *all* of the following subjects relating to content knowledge: physics (lecture [4 credits] and practical work [2 credits] involving ICT), chemistry (lecture [4 credits] and practical work [2 credits] involving ICT), biology (lecture [4 credits] and practical work [2 credits] involving ICT), and earth science (lecture [4 credits] and practical work [2 credits] involving ICT). These subjects fall under the category of "teaching subjects" and are intended for teaching certificates sought by lower and upper secondary school science teachers. However, as mentioned above, the minimum requirement of "teaching subjects," which has the same meaning as "content knowledge" has been drastically reduced since the millennium. This fact signifies that Japanese teachers are required to develop practical competencies.

Related to subjects on science education are subjects grouped with "methods of subject teaching" (Table 14.3). The subject "general principles and methods of science teaching" (4 credits) is mandatory for teaching certificates of science, which include knowing *why* and *what* to teach science, understanding *how* to teach science, and assessing students' performance based on research results and real problems in science lessons. The Faculty of Education at Hiroshima University also provides another compulsory subject "science curriculum" which includes the organizing and management of the science curriculum, and knowing and understanding the course of study, and also delivers a variety of specialized subjects relating to science education as elective. Among these subjects are "theories of assessment and its methods in science," "history of science teaching," "science education in the world," and "seminar on science teaching" that focuses on PCK through researching and developing teaching materials, and making lesson plans. In contrast, while science faculties which are designated as teacher training institutions by the MEXT, generally provide "the methods of subject teaching" as far as minimum requirement is concerned, they also provide a variety of "teaching subjects" as part of their identity. This work involves teaching the minimum requirement of four science subjects (lecture and practical work involving ICT in physics, chemistry, biology, and earth science). Clearly, there is still a debate regarding the professionalism and academicism models of secondary teachers that occurred before World War II.

From the perspective of didactic transposition theory (Chevallard, 1989; Bosch and Gascón, 2006), *scholarly knowledge* (scientific knowledge), in the research context, cannot be directly transferred into the minds of students. Science teachers have to transpose *scholarly knowledge* into teachable and learnable forms called *knowledge to be taught* in the societal context (e.g., curriculum, textbooks), and then *taught knowledge* in the classroom, while keeping its power and functional character. For science teachers who graduated from either educational or scientific faculties, the important thing is to know the characteristics of the types of knowledge and link them to pedagogical approaches for teaching science.

According to a research on beliefs about the aims and objectives of science teaching for experienced science teachers (Ueda & Isozaki, 2016), pre-service teacher training should play a vital role in providing crucial opportunities for the development of beliefs about aims and objectives of science teaching. Therefore, a variety of subjects on science education should be offered together with subject studies and pedagogical studies in pre-service teacher training.

A System of Teaching Practice

In many universities of education or educational faculties, a variety of teaching practice is provided. The Faculty of Education at Hiroshima University provides a series of step-by-step teaching practices from the first to the third year. These teaching practices occur in the secondary schools attached to Hiroshima University. During the introductory stage of teaching practice, the focus is on the transformation from a student's view to a teacher's view of a science lesson through observation of science classes. The second year focuses on recognizing how student teachers engage in lesson study by observing student teachers' behavior. Finally, at the third year, there are two types of teaching practices: the first involves observing their mentor's work for three days before the four-week teaching practice, and the second is engaging in lesson study themselves under the instruction of mentors for four weeks (two weeks in school A and another two weeks in school B). During the four-week teaching practice, student teachers also participate in other activities in both secondary schools. Teaching practice should be situated in an environment where student teachers can bridge the gap between theories and practice under the supervision of mentors within the professional community.

Lesson Study and Pedagogical Content Knowledge

In Japan, lesson study is a traditional art of investigating teaching and learning, and engaging with colleagues in Japan. It can typically be divided into three parts: preparation, research lesson, and reflective meeting/conference. Student teachers are encouraged to participate in teacher-related activities within a professional community that includes experienced teachers (mentors), other student teachers, and the school students themselves. The process focuses on various dimensions: what teaching involves; what the teaching profession is like in actual practice; how to carry out lessons under the instruction of a mentor and a university tutor; how to make a lesson plan and a scheme of work; and how to research and develop teaching materials individually and/or with other student teachers under the supervision of mentors. Student teachers observe their mentor's model lessons and attend lectures where mentors share their view on teaching and classroom management. Student teachers also take part in a reflective meeting/conference with their peers after a research lesson is given by one of the student teachers. This process is repeated throughout the students' teaching practice (Isozaki, 2015).

PCK is quite a complicated term and has been given various interpretations by researchers (e.g., Gess-Newsome and Lederman, 1999; Bishop and Denley, 2007). One argument is that it can be developed during teaching practice, especially in lesson study. PCK can play an important role in the lesson study preparation phase, especially in researching and developing teaching materials, and making lesson plans (Isozaki, 2015). This is because teachers must transform their content knowledge for it to become teachable and learnable knowledge. Furthermore, teaching involves a series of professional judgements and decisions under the complex contexts in practice. Student teachers as well as teachers toned to think and choose the best plan and method in order for students to learn. Figure 14.2 shows a typical lesson plan in Japan. According to Shulman's model of pedagogical reasoning and action (Shulman 1987, p. 16), this transformation would entail the following processes: (a) *preparation*, including critical interpretation; (b) *representation* of the ideas as analogies, metaphors, etc.; (c) instructional *selections* of teaching methods and models; and (d) *adaptation* to the general characteristics of students; as well as (e) *tailoring* these adaptations to specific students. In transformation, student teachers are required to make crucial judgement and decision using their knowledge base, especially PCK

Grade ○ Science Lesson Plan

Name of Teacher:
Date and Time: yy/mm/dd, time
Place: laboratory or other places
Class:

【Unit title】

【Unit objectives】based on the Course of Study, and teacher's own ideas: Student should know and understand, be able to, etc.

【The views of unit】

1) Main ideas of the unit, and the key teaching contents,
2) Learners' characteristics, their prior knowledge and preconceptions relating to this unit, and classroom atmosphere, and
3) Teacher's view and ideas for instruction, based on both 1) and 2).

【Assessment task】 criteria and methods of the whole unit and each lesson.

【Scheme of work】 the sequence of unit goals with the number of hours to be spent on the unit as well as a description of what might happen during the one-hour lesson.

【Today's lesson】

1) Objectives of today's lesson
2) Title of topic
3) Resources
4) Development of today's lesson as follows:

Time & Phase	Students' activities and learning based on *anticipated* students' response	Teacher's activities involving assessment task, and notes
(minutes) [Introduction] [Development] [Conclusion]	[review of previous lesson] [some questions and *anticipated* answers] [practical activities and *anticipated* errors] [students' presentation and *anticipated* results] etc.	[points to keep in mind] [assessment tasks: criteria and methods including supports for lower achieved students] [safety] etc.

Figure 14.2 A typical lesson plan in Japan.
Source: Isozaki, T. (2015); partially modified by the authors.

with advice from mentors. In other words, student teachers need to anticipate students' response, know their students, select appropriate teaching materials, understand the teaching context, and integrate all this information in developing a lesson plan (Figure 14.2). As a result, lesson study can offer a combination of theoretical knowledge with practical know-how. This transformation helps develop PCK, and through this process, student teachers gradually become familiar with the processes and benefits of lesson study as a traditional professional culture.

According to Ochi and Isozaki (2016), student teachers' knowledge base is acquired/developed through the following ways: capturing their own classroom practice, engaging in reflective meeting/conference, and observing other student teachers based on their view of (science) lessons. Further observations, supervision by mentors, and collaboration with colleagues will enable student teachers to conduct teaching practice more reflectively.

Although the period of teaching practice is quite short compared to other western countries such as the United Kingdom and Finland, teaching practice is the important opportunity to be mentored by experienced teachers to learn the explicit and hidden norms of a professional community and a professional culture. Therefore, teaching practice in pre-service teacher training can be situated in the introduction stage of continuing professional development.

Opportunities to Learn the Way of Learning From Other Field Experiences

It is worthwhile to note that university students can gain opportunities for formal and informal teaching experiences. Since master course students have already obtained teacher certificates (class I) in science when they graduated from bachelor courses, they can work as part-time science teachers at lower or upper secondary schools with the same responsibilities for teaching science without other activities as full-time science teachers. It is a very useful field experience opportunity for them to become a science teacher. Some university students can volunteer in taking care of children at community schools that are supported by the board of education, parents, local peoples, and others stakeholders. Traditionally some university students teach science or other subjects at *jyuku* (cram school) two or three evenings per week as a part-time job.

These formal and informal opportunities of teaching science can accelerate enthusiasm to become a science teacher, provide avenues to observe and recognize children's behavior, and increase awareness about the difficulties in teaching science.

BECOMING A SCIENCE TEACHER

To be a secondary science teacher for a municipal school in Japan, a person must pass an appointment/hiring examination managed by the board of education. The appointment examination is conducted with a combination of various selection methods to evaluate knowledge, competencies, attitudes, and other teaching skills. There are written examinations in science (whole area and/or specialized area), general literacy and pedagogical studies. There are also interviews (individual and/or group), essay examination, preparation of lesson, and short time trial lesson in front of examiners who are consultant teachers of the board of education, principals and others, aptitude tests, and so on. The methods of appointment examination depend on respective board of education. In general, the examinations are divided into two or more steps. The first step is an examination held in July for the next fiscal year. In Japan, an academic year starts in April. The examinees who pass the first step examination can take the second or final examination. Even if candidates pass the examinations, they are not automatically appointed as a teacher. It means that the candidates are registered on a list based on their examination scores and will be employed according to their rank on the list.

There are two routes to be a science teacher at a national school that is attached to a national university: the first is to be directly employed by the national school itself, the second is to exchange a teacher between a national school and a municipal school. The latter route limits working years at a national school.

Table 14.4a and b shows the applicants and successful candidates in two cases, providing some vital information. Table 14.4a shows that the appointment examination of Tokyo is conducted by category for secondary school level. In Osaka (Table 14.4b), the examination is conducted by category for each school level. This means that a science teacher will be able to move between lower and upper secondary schools in Tokyo, but they need to have both teaching certificates for lower and upper secondary schools. On the other hand, in the case of Osaka, a science teacher who is employed as a lower secondary school teacher will spend his/her career life in same school level. Second, the number of applicants, examinees, and successful candidates of biology and chemistry are much higher compared to physics and earth science. These facts reflect the real condition of science education at upper secondary schools. Chemistry and biology are popular for both science and non-science students. However, earth science is the most unpopular among science subjects. Therefore, many upper secondary schools do not have earth science specialists. As a result, earth science has not been offered as an elective subject in many schools ever since the selective system of science subjects started in the 1970s (Isozaki, 1996). Teachers

TABLE 14.4a The Result of Appointment Examinations by Tokyo Metropolitan Board of Education in 2015 (Both Lower and Upper Secondary Schools)

	No. of Applicants	No. of Examinees (a)	No. of Registered on A List (b)	Examination Magnification (a/b)
Physics	249	209	31	6.7
Chemistry	318	275	40	6.9
Biology	347	299	53	5.6
Earth Science	44	39	1	39.0

Source: Based on Tokyo Metropolitan Board of Education (2015). *The result of appointment examinations for the fiscal 2016* (in Japanese). Calculated by the authors. Retrieved from http://www.kyoiku.metro.tokyo.jp/press/2015/pr151016/besshi.pdf.

TABLE 14.4b The Result of Appointment Examinations by Osaka Prefecture Board of Education in 2015 (There are Two-Steps Examinations)

	No. of the First Examination Applicants	No. of Examinees of the First Examination (a)	No. of the Final Successful Candidates (b)	Examination Magnification (a/b)
Lower Secondary School Science	318 (53)	199 (48)	71 (29)	2.8
UP Physics	123 (15)	77 (14)	16 (5)	4.8
UP Chemistry	170 (22)	103 (19)	23 (9)	4.5
UP Biology	140 (18)	90 (16)	26 (10)	3.5
UP Earth Science	17 (2)	14 (2)	6 (1)	2.3

Notes: UP means upper secondary school. The number in parenthesis is the number of the first examination exempt.

Source: Based on Osaka prefecture Board of Education (2015). *The result of the second appointment examination for the fiscal 2016* (in Japanese). Calculated by the authors. Retrieved from http://www.pref.osaka.lg.jp/kyoshokuin/kyosai/h28_2_kekkahyo.html

specializing in other fields such as physics are assigned to teach earth science instead because their teaching certificate in "science" signifies that they have studied earth science and its practical work.

FUTURE CHALLENGES

Teaching has been recognized as a profession that is associated with human relationships and highly sophisticated intellectual competencies since the late 1990s (e.g., Teacher Education Council, 1997). As a result, a master's level course for teacher training has been offered within postgraduate schools in many universities. For a professional school that aims to develop and enhance practical competencies (not enhancing teaching subject) in teaching, the new graduate school for a master's course was introduced in 2008. The Central Council for Education (*Chuō kyouiku shingikai,* in Japanese, 2012) criticized the division of teacher education between pre-service by universities and in-service by boards of education (see Figure 14.1). It was recommended that a unified system of teacher education based on collaboration and cooperation between universities and boards of education should be constructed in order to ensure continuing professional development in the teaching profession. As these reforms illustrate, Japan's teacher education has undergone innovations since the millennium. There is a consistent thrust to enhance the teaching profession, and develop a teacher's model of continuing professional development. However, there remain old and new controversial issues in teacher education about the ideal secondary science teacher.

Ogawa (2014) and Isozaki (2016) effectively described the science teacher's culture. According to them, it is important that Japanese science teacher's traditionally accumulated wisdom and expertise based on their enthusiasm and reflective practices as a professional culture (e.g., lesson study based on collaboration with colleagues), should be reflected in the innovations in science teacher education in Japan.

REFERENCES

Aldrich, R. (1996). *Education for the nation.* London, England: Cassell.

Bishop, K., & Denley, P. (2007). *Learning science teaching: Developing a professional knowledge base.* Maidenhead, England: Open University Press.

Bosch, M., & Gascón, J. (2006). Twenty-five years of the didactic transposition. *ICMI Bulletin, 58,* 51–63.

Central Council of Education. (2012). *Report on strategies to improve teacher's competencies through continuing professional development: A proposal.* Tokyo, Japan: MEXT. Retrieved from http://www.mext.go.jp/component/b_menu/shingi/toushin/__icsFiles/afieldfile/2012/08/30/1325094_1.pdf [in Japanese]

Chellavard, Y. (1989). On didactic transposition theory: Some introductory notes. In *Proceedings of the international symposium on selected domains of research and development in mathematics education* (pp. 51–62). Bratislava, Slovakia. Retrieved from http://yves.chevallard.free.fr/spip/spip/IMG/pdf/On_Didactic_Transposition_Theory.pdf

Gess-Newsome, J., & Lederman, N. G. (Eds.). (1999). *Examining pedagogical content knowledge: The construction and its implementation for science education.* Dordrecht, The Netherlands: Kluwer Academic.

Isozaki, T. (1996). A survey of earth science education in Japan. In D. A. V. Stow & G. J. H. McCall (Eds.), *Geoscience education and training: In schools and universities, for industry and public awareness,* (pp. 93–107). Rotterdam, The Netherlands: A. A. Balekema.

Isozaki, T. (2001). History of science teacher education: "What was the professional knowledge and competence of science teachers before World War II?" *Journal of Science Education in Japan, 25*(1), 11–23. [In Japanese with English abstract]

Isozaki, T. (2015). Lesson Study research and practice in classroom. In R. Gunstone (Ed.), *Encyclopedia of Science Education* (pp. 615–618). Dordrecht, The Netherlands: Springer.

Isozaki, T. (2016). How have Japanese *Rika* (school science) teachers traditionally formed their own cultures and improved their teaching competencies through research and practice? In Mei-Hung Chiu (Ed.), *Science education research and practice—Challenges and opportunities* (pp. 517–537). Dordrecht, The Netherlands: Springer.

Isozaki, T., & Isozaki, T. (2011). Why do teachers as a profession engage in lesson study as an essential part of their continuing professional development in Japan? *International Journal of Curriculum Development and Practice, 13*(1), 31–40.

Iwata, Y. (2006, October). *Teacher education and neo-liberalism—Japan and other countries.* Key presentation at round-table session at 2nd International Conference of Teacher Education, East China Normal University, Shanghai China. Retrieved from https://www.u-gakugei.ac.jp/~currict/about/iwata.info/20061027shanghai.pdf

Kumura, T., & Iwahashi, B. (1967). Development of the teacher training system in Japan. *Education in Japan, II,* 75–89.

Ministry of Education, Culture, Sports, Science, and Technology. (2013). *2012 White Paper on Education, Culture, Sports, Science and Technology.* Tokyo, Japan: MEXT. Retrieved from http://www.mext.go.jp/b_menu/hakusho/html/hpab201201/detail/1345171.htm

Ministry of Education, Culture, Sports, Science, and Technology (2015). *Universities that provide teacher certificates in 2015.* Tokyo, Japan: MEXT. Retrieved from http://www.mext.go.jp/component/a_menu/education/detail/__icsFiles/afieldfile/2015/12/01/1287058_1.pdf. (in Japanese).

Ministry of Education, Culture, Sports, Science, and Technology (2016). *School statistics in 2016.* Tokyo, Japan: MEXT. Retrieved from http://www.mext.go.jp/b_menu/toukei/002/002b/1368900.htm. [1368897_10.xls] [in Japanese with English terms]

Ministry of Education, Culture, Sports, Science, and Technology (n.d.). *Education personnel certificate action and regulations.* Tokyo, Japan: MEXT. Retrieved from http://www.mext.go.jp/a_menu/koutou/kyoin/1268593.htm. [in Japanese]

Ministry of Education, Science, and Culture. (1980). *Japan's modern educational system.* Tokyo, Japan: Ministry of Finance.

Ochi, T., & Isozaki, T. (2016). How do pre-service science teachers develop their teacher knowledge? A qualitative study focusing on teaching practice in schools. *Theory and Research for Developing Learning Systems, 2,* 23–33.

Ogawa, M. (2014). Occupational culture as a means of professional development for preservice science teachers in Japan. In Chen-Y. Lin & Ru-J. Wang (Eds.), *Innovations in science teacher education in the Asia Pacific* (pp. 61–80). Bingley, England: Emerald.

Ogawa, M. (2015). Rika. In R. Gunstone (Ed.), *Encyclopedia of Science Education* (p. 840). Dordrecht, The Netherlands: Springer.

Organization for Economic Cooperation and Development. (n.d.). *Result from TALIS 2013 Country note: Japan.* Paris, France: Author. Retrieved from https://www.oecd.org/japan/TALIS-2013-country-note-Japan.pdf

Osaka Prefecture Board of Education (2015). *The result of the second appointment examination for the fiscal 2016.* Osaka, Japan: Author. [in Japanese] Retrieved from http://www.pref.osaka.lg.jp/kyoshokuin/kyosai/h28_2_kekkahyo.html

Peters, R. S. (1966). *Ethics and education.* London, England: George Allen & Unwin.

Shimahara, N. (2002). *Teaching in Japan: A cultural perspective.* New York, NY: Routledge.

Shulman, L. S. (1987). Knowledge and teaching: Foundations of the new reform. *Harvard Educational Review, 57*(1), 1–22.

Teacher Education Council. (1997). *Report on strategies to improve teacher training for a new age: The first proposal.* Tokyo, Japan: MEXT. [in Japanese]

Tokyo Metropolitan Board of Education (2015). *The result of appointment examinations for the fiscal 2016.* Tokyo, Japan: Author. Retrieved from http://www.kyoiku.metro.tokyo.jp/press/2015/pr151016/besshi.pdf [in Japanese]

Ueda, Y., & Isozaki, T. (2016). Research into development of beliefs about the goals or purposes of science teaching: Analysis of life stories of five experienced science teachers. *Theory and Research for Developing Learning Systems, 2,* 35–47.

CHAPTER 15

INTERNATIONAL COMPARISON OF SCIENCE TEACHER PREPARATION

Various Challenges in Different Contexts

Toshihide Hirano
Aichi University of Education

Jon E. Pedersen
University of South Carolina

The goal of this volume was to provide the reader an opportunity to engage in understanding various countries' approaches to science teacher preparation. As we mentioned in the introduction, we certainly will not claim that we have covered the globe in terms of all the countries that could have, or maybe should have, contributed to this volume. What we will claim is that we have a very good sample of countries that offer distinct and varying perspectives on how to best prepare science teachers. In the process of writing these chapters, authors were specifically asked to address both the strengths and weaknesses in their countries' approaches. What we are attempting to

Model Science Teacher Preparation Programs, pages 307–325

accomplish in the final chapter of this book is an internal comparison of all these various perspectives.

We certainly recognize the numerous challenges represented in the varying contexts of 14 countries, especially in the descriptions of initial teacher training programs for secondary school science. The environment of science education and science teacher education varies widely among those countries participating regarding government budget, the diffusion or promotion of reform for secondary education, and the nation's interest in the teaching profession. More specifically, these challenges have shaped the actions of each country's government, educational institutions, schools, teachers, teacher educators, and prospective science teachers as they attempt to address the gap between educational theories and practices, aims and results, demand and supply, and so on. Within this final chapter we summarize and share not only the challenges but some of the solutions different countries have offered that address concerns that cut across countries.

All of the data offered from each chapter were analyzed comparatively based on the guiding framework initially established for this volume to better understand each country's perspective and to provide for the reader an interpretive overview in a manner in which to compare science teacher preparation as represented by the participating countries. The results of that analysis is summarized into three areas as follows:

1. System of the initial teacher education.
2. Training program requirements.
3. Strength and improvements of science teacher preparation.

SYSTEM OF THE INITIAL TEACHER EDUCATION

The characteristics of the initial teacher education (ITE) system of 14 countries are shown in Table 16.1a–Table 16.1c: divided into three parts by the international area classification. The ITE program framework is often prescribed at a national level, or entrusted to a local government level in countries such as Canada, Germany, United Kingdom, and the United States. Australia is shifting to a national level framework by introducing the nationwide accreditation system. In Europe, the evidence of ITE program reform implemented by the Bologna Process could be found in each member country. From the contents described in these three tables, the findings of a multi-country comparison are:

- The frame of ITE curriculum is established mainly by laws, ordinances, or standards by the government. But in some countries like Australia, Germany, and the United States, it depends on the accreditation system framework for teacher training (see for example Council for Accreditation of Professional Education [CAEP]).

- The undergraduate-level teacher education courses at universities predominates in ITE, but in Europe it is developed at a master level. Not only are 4-year BEd (Bachelor of Education) courses common, but also 1- or 2-year teacher training course for BSc (Bachelor of Science) graduates are common.
- As a qualification for entering ITE programs, it is common for most countries to impose an examination for applicants checking their literacy, competency, upper secondary school graduate qualification, university degree qualification, health, or aptitude.
- In the case where the science teaching certificate is based on specialized subjects like physics, chemistry, and biology, the qualification requirement of content knowledge (CK) is usually in one subject area (major), or two subject areas (double major or major/minor). But for a general science or integrated science certificate, students are required to take each subject areas basic courses. For most countries, there are minimum requirements of CK for upper secondary school science certification where students are focusing on a single subject (e.g., physics), or whole science subjects like Japan.
- In Australia, Germany, Israel, United Kingdom, and the United States, the demonstration of, or record of presentations of science lessons in practice meeting specific country (or local) standards, is required as a qualification for completing the ITE program. The teacher selection examination in Japan, Korea, and Turkey which comes after finishing the ITE program becomes the the qualification for teaching.
- Science teachers' employment in many countries (particularly physics teachers) tends to be short. In the United Kingdom, school-based training for non-salaried or salaried individuals is introduced as an alternative route for university-based training and special scholarships are provided for teacher trainees (the amount of the scholarship changes according to level of need for specific subject area teachers). In the United States, alternative certificate programs exist both inside and outside of the university and teacher employment programs like Teach for America (TFA) and Troops to Teachers (TTT) are available for scholars who meet specific qualifications.
- There are two ways of implementing student teaching (or practicums) and internships. One is setting several times during the program period for students to engage with students in the schools. The other is setting a single time in the final year of the preparation experience. Israel, Japan, Korea, and Turkey are the exceptions to

TABLE 16.1a International Comparison of the Systems of Initial Teacher Education Between 14 Countries (Part 1: Europe)

		Europe				
		Portugal	France	U.K. (England)	Germany	Finland
System of the Initial Teacher Education	Forming ITE framework (regulation law, accredition, standards etc.)	National: regulated by a Decree Law 79/2014	National: National Standards of Teachers' Competencies (MoE, 2013)	National: QTS (NCTL), The Teachers' Standards (DfE, 2012), Ofsted inspection framework	State: Program & system accreditation by 10 institutes, common standards of the KMK (2004, 2008)	National: regulated by law, recommendations, and agreement
	Secondary science subjects	physics and chemistry, biology and geology	physics and chemistry, biology and geology	physics, chemistry, biology, physics with mathematics	physics, chemistry, biology	mathematics, physics, chemistry, biology, geology
	CK requirements: 1 specialist subject			✓		✓(major)
	CK requirements: more than 2 subjects	✓(2 subjects)	✓(2 subjects)	✓for lower secondary teacher	✓2 subjects	✓other 1–2 subjects (minor)
	University-based ITE route	5 years (300 ECTS) at higher education institution (*1st cycle for disciplines)	2 years at ESPE (*after completing BSc)	3–4 years at univ., or 1–2 years at PGCE for graduates (60 days at Univ.)	5 years (300 ECTS) BSc–MEd at univ. (*taking two subject areas	5 years (300 ECTS) at univ. (*taking two subject areas)
	Degree for the certification	Master	Bachelor and Master	Bachelor	Bachelor and Master	Bachelor and Master
	Start on taking pedagogical studies	Master (2nd cycle: 2 years)	Master MEEF or after completing degree	Undergraduate or Post-Graduate	Master	Undergraduate or after completing degrees

(continued)

TABLE 16.1a International Comparison of the Systems of Initial Teacher Education Between 14 Countries (Part 1: Europe) (continued)

		Europe				
		Portugal	France	U.K. (England)	Germany	Finland
System of the Initial Teacher Education	Qualification for entering ITE program or field experience		Examination: (M1 MEEF) CAPES, (Master's degree) Agregation	Grade score and QTS Skills Test	(Abitur)	Test: Interviews by the staffs of the university and training school
	Qualification for completing ITE			report practices met the Standards	examination lesson (one lesson for each subject), oral examination, and a thesis	
	Alternative ITE route			School-based training route: SD non-salaried (60 days), SD salaried (16 days), Teach First and SCIIT(6 days)		
	Patterns and features of school experience	Introduction to professional practice (48 ECTS, 1 year): – 4–6 weeks supervised teaching practice – research project	Internship (20 ECTS): – coaching by field and academic tutors – recommend 3 types of internship: observation, practice, responsibility	PGCE & SD non-salaried/24 weeks in two years: – at two different schools – judged by 6 observed lessons using Ofsted criteria	In-service training (from 18 months to 2 years): – feedback from regular teacher and teacher educator – final examination lesson	Supervised teaching practice (20 ECTS, 2–3 years): – univ. attached training schools with experienced mentor teachers

TABLE 15.1b International Comparison of the Systems of Initial Teacher Education Between 14 Countries (Part 2: Middle East, North and Latin America)

		Middle East		North and Latin America		
		Turkey	Israel	Canada	U.S.A.	Argentina
System of the Initial Teacher Education	Forming ITE framework (reguration law, accredition, standards etc.)	National: common curriculum designed by Higher Education Council	National: academic and practical standards by MOE	Province: accreditation by provincial MOE or Professional organization	State: CAEP accreditation, NSTA Standards for Science Teacher Preparation (NSTA, 2012)	National: national curricular guidelines for teacher education (Federal Board of Education, 2007)
	Secondary science subjects	science (4 areas), physics, chemistry, biology	general science, physics, chemistry, biology, earth sciences	general science, physics, chemistry, biology	general science, physics, chemistry, biology	physics, chemistry, biology, natural sciences
	CK requirements: 1 specialist subject	✓	✓		✓	✓
	CK requirements: more than 2 subjects	✓for Science teacher	✓for general science teacher	✓2 subjects	✓for general science teacher	✓for natural science teacher
	University-based ITE route	Science: 4 years at univ. Upper Secondary Subjects: 2-semesters PFCP for graduates	4-year BEd, or 2-year external program from 3rd-year BSc	4-year BEd, 16-month post-graduate, or 5-year BSc+BEd	4-year at univ., or 1-year post-graduate	4-year at teacher education institutes and univ.
	Degree for the certification	Bachelor	Bachelor	Bachelor	Bachelor	Bachelor

(continued)

TABLE 15.1b International Comparison of the Systems of Initial Teacher Education Between 14 Countries (Part 2: Middle East, North and Latin America) (continued)

		Middle East		North and Latin America		
		Turkey	Israel	Canada	U.S.A.	Argentina
System of the Initial Teacher Education	Start on taking pedagogical studies	undergraduate or post-graduate	undergraduate	undergraduate or post-graduate	undergraduate or post-graduate	undergraduate
	Qualification for entering ITE program or field experience	university entrance exam: YGS and LYS		requirement: grade score, completion of science courses, and exceeding the GPA	States exam: (enter) literacy skills, GPA, interview/(clinical experience) CK, GPA, teaching skills	(some universities) health exam, attend introductory course
	Qualification for completing ITE	(Public personnel selection exam: KPSS)	Evaluation: observation of class session by school principle & MoE inspector		final clinical experience practice portfolio	
	Alternative ITE route				alternative certification program used by states, TFA and TTT (no pedagogical knowledge teacher)	
	Patterns and features of school experience	Science: school experience (5 hours), teaching practice(8 hours) Upper Secondary Subjects: teaching practice (5 credits)	school-based practice of 90–180 hours	field experiences (more than 16 weeks): – combination of observation, orientation, independent teaching, community service	early field placement (15–30 hours)/ methods course/ student teaching (more than 10 weeks full-time) – required performance-based assessment	professional exercise and teaching practice I–IV (every year): – focus on understanding educational contexts, schools, classroom, and teaching

TABLE 15.1c International Comparison of the Systems of Initial Teacher Education Between 14 Countries (Part 3: Asia and Pacific)

		Asia and Pacific			
		Australia	China	Korea	Japan
System of the Initial Teacher Education	Forming ITE framework (reguration law, accredition, standards etc.)	National: accreditation standards and procedures policy (AITSL), Australian Professional Standards for Teachers (AITSL,2011)	National: The Secondary School Teacher's Professional Standards (MoE, 2012 trial)	National: qualification criteria without national assessment	National: approval system, and accreditation criteria
	Secondary science subjects	general science, physics, chemistry, biology	physics, chemistry, biology, integrated science, technology	common sciences, physics, chemistry, biology, earth science	general science, physics, chemistry, biology, earth science, integrated science
	CK requirements: 1 specialist subject	✓(Major)	✓	✓	(additing as a major area)
	CK requirements: more than 2 subjects	✓other 1 subject (miner)	✓for Integrated Science teacher	✓for Common Sciences teacher (second-major)	✓
	University-based ITE route	4-year BEd at univ., 3-year BSc + 1-year DipEd or 2-year MTeach at Univ.	4 years at univ. or teacher college	4-year BEd at univ., or BSc + MEd	4 years at univ. with certified teacher-training course
	Degree for the certification	Bachelor or Master	Bachelor (Master: advance level)	Bachelor or Master	Bachelor (Master: advance level)
	Start on taking pedagogical studies	undergraduate or post-graduate	undergraduate	undergraduate or post-graduate	undergraduate

(continued)

TABLE 15.1c International Comparison of the Systems of Initial Teacher Education Between 14 Countries (Part 3: Asia and Pacific) (continued)

		Asia and Pacific			
		Australia	China	Korea	Japan
System of the Initial Teacher Education	Qualification for entering ITE program or field experience	university decided (TEMAG recommends using academic achievement)	teacher qualification certification exam (MOE, starting 2015)	(undergraduate) the personel selection for the entrance quota of 10%	(university entrance exam)
	Qualification for completing ITE	University decided (TEMAG recommends including final assessments of evidence portfolio)		Average grade, test of teaching aptitude and personality (appointment exam EASST)	(Prefectual public school teacher selection test for employment)
	Alternative ITE route				
	Patterns and features of school experience	practicum: 3 weeks each year/semester (Bachelor 3 years/MTeach 3 semesters) educational internship: 10 weeks (last year/semester) – Teaching 20 hours/ week	Practicum I: – Observe teaching and management and report Practicum II (12 weeks): – Guide by observed univ. Faculty member	school internship (4–6 weeks): – Pre-guidance, visit meeting, interview and survey by univ. Educational service activities (more than 30 hours)	Junior high (3–4 weeks)/ senior high (2 weeks): – Pre & post meeting with university professors – Lesson study and daily journal writing for reflection

this where their teacher trainees are performing in a school for more than ten weeks.

- In the United Kingdom, it is necessary for student teaching to be planned and carried out at two different schools. However, there is disagreement in the United Kingdom (and other countries) about the length and times of the training period, and number of practice lessons that each teacher trainee must complete at the school.
- In China, Finland, France, Germany, Japan, Korea, and the United States, university staff members and school teachers focus on constructing good relations with each other in order to guide the teacher trainee jointly. Moreover, the teacher trainees in France and Portugal carry out research activities at their training school during the student teaching period.

TRAINING PROGRAM REQUIREMENTS

Table 16.2a and Table 16.2b show the content requirements of 14 countries' university-based training programs, that are divided into two parts by the international area classification. In the upper part of the tables, there are the calculated component ratios of each course for CK, general pedagogical knowledge (GPK), pedagogical content knowledge (PCK), and field experience (FE) in each country's training program based on the description of courses' credit hours by chapter authors. Except for the Postgraduate Certificate in Education (PGCE) program in the United Kingdom (which is for BSc graduates and not accompanied by the setting of CK subjects), most university-based training programs are 4-year undergraduate programs of each country. In the middle and lower part of the tables, there are two checklists showing the content components of the general pedagogy courses (which are related to GPK) and subject teaching courses (which are related to PCK mainly, but also some CK) of each training program made from the description of course subject names by chapter authors. From the contents described in these two tables, the findings of a multi-country comparison are stated below:

- Among 14 countries, the mean image of content ratios for CK, GPK, PCK, and FE are 60%, 20%, 10%, and 10% in round numbers. PCK may sometimes be bound with CK or GPK.
- To compare with the mean image, GPK is lower and CK is higher in China; CK is lower and GPK is higher in Australia, Japan, Turkey, and the United States; and GPK is lower and PCK and FE are higher in France and Portugal.
- Among 14 countries, the common components of general pedagogy courses are "basics of pedagogy, planning, and teaching" and "ba-

TABLE 15.2a International Comparison of the Training Program Content Requirements Between 14 Countries (Part 1: Europe, Middle East)

		Europe					Middle East	
		Portugal	France	U.K. (England)	Germany	Finland	Turkey	Israel
University-Based Route Program	Component Ratio: CK	50%	32%	(PGCE)	66%	66%	49%	57%
	Component Ratio: GPK	8%	10%	33%	12%	20%	30%	21%
	Component Raio: PCK	22%	39%		16%	7%	10%	10%
	Component Ratio: Field Experience	20%	19%	67%	6%	7%	11%	12%
Contents of General Pedagogy Course	Basics of pedagogy, planning and teaching	✓	✓	✓	✓	✓	✓	✓
	Basics of psychology, understanding pupils	✓	✓	✓	✓	✓	✓	✓
	Classroom mannagement		✓	✓	✓	✓	✓	
	Assessment & evaluation		✓	✓	✓	✓		✓
	Professional attitudes	✓	✓	✓		✓		
	Special needs & additional foreign language		✓	✓				
	Educational research methods	✓	✓					
Contents of Subject Teaching Course	Science teaching and learning strategies	✓	✓	✓	✓	✓	✓	✓
	Pupils' understanding of science	✓	✓	✓	✓	✓	✓	✓
	Science curriculum knowledge	✓	✓	✓	✓	✓	✓	
	Scientific process and method	✓	✓	✓	✓	✓	✓	
	Assessment of scientific literacy	✓			✓	✓		✓
	History of science		✓			✓	✓	
	Applications and implications of science		✓	✓	✓		✓	
	Research & development of science education		✓		✓	✓		
	Purposes for teaching science	✓			✓	✓		
	Experiment operations and management				✓			

TABLE 15.2b International Comparison of the Training Program Content Requirements Between 14 Countries (Part 2: North and Latin America, Asia and Pacific)

		North and Latin America			Asia and Pacific			
		Canada	United States	Argentina	Australia	China	Korea	Japan
University-Based Route Program	Component Ratio: CK	60%	48%	50–60% (incl.PCK)	38%	73%	58%	31%
	Component Ratio: GPK	29%	35% (incl. PCK)	25–30%	38%	9%	25%	49%
	Component Raio: PCK	11%	1 course	2–3 courses	12%	18%	11%	12%
	Component Ratio: Field Experience	16 weeks	18%	15–25%	12%		6%	8%
Contents of General Pedagogy Course	Basics of pedagogy, planning and teaching	✓	✓	✓	✓	✓	✓	✓
	Basics of psychology, understanding pupils	✓	✓	✓	✓	✓	✓	✓
	Classroom management	✓	✓		✓	✓	✓	✓
	Assessment & evaluation	✓	✓		✓		✓	✓
	Professional attitudes			✓		✓	✓	✓
	Special needs & additional foreign language	✓	✓		✓		✓	✓
	Educational research methods					✓		
Contents of Subject Teaching Course	Science teaching and learning strategies	✓	✓	✓	✓	✓	✓	✓
	Pupils' understanding of science	✓			✓	✓	✓	✓
	Science curriculum knowledge	✓	✓		✓	✓	✓	✓
	Scientific process and method					✓	✓	
	Assessment of scientific literacy					✓	✓	✓
	History of science			✓		✓	✓	
	Applications and implications of science			✓				
	Research & development of science education				✓ (MTeach)			
	Purposes for teaching science							✓
	Experiment operations and management						✓	✓

sics of psychology, understanding pupils," and that of subject teaching courses is "science teaching and learning strategies." Those are very traditional, theoretical, and essential components teacher educators could not omit.

- The components prepared in more than 10 countries are "classroom management" and "assessment and evaluation" in general pedagogy courses, and "pupils' understanding of science" and "science curriculum knowledge" in subject teaching courses. Those are also theoretical components, but needing more practical training to acquire the instructional skills.
- In general pedagogy courses, the components of "professional attitudes" and "special needs and additional foreign language" are prepared in more than seven countries. In China, France, and Portugal, there is a component of "educational research methods." Those components are needed to ensure understanding and implementing the universal design for learning in education.
- In subject teaching courses, more than seven countries prepare the components of "scientific process and methods" and "assessment of scientific literacy." And more than five countries prepare the components of "history of science," "applications and implications of science," and "research and development of science education." Also, the component "purposes for teaching science" in Finland, Germany, Japan, and Portugal; and "experiment operations and management" in Germany, Japan, and Korea, are prepared. Those components are needed to ensure understanding and implementing the student-centered and practice-based authentic science for pupils.
- In the Europe region and the Asia-Pacific region, there are many course content components related to GPK and PCK than in other regions.

STRENGTH AND IMPROVEMENTS OF SCIENCE TEACHER PREPARATION

In each chapter, authors provided their review for the present conditions of their country's initial teacher training program for secondary school science, which suggests the strength and areas for improvements of science teacher preparation. The main claims of 14 countries relating to the category of CK, GPK, PCK, and others are shown in Table 16.3a–Table 16.3c divided into three parts by the international area classification. The indications of each category are as follows:

- CK: As CK occupies the majority of the teacher knowledge, it is thought that high quality and enough quantity of CK is necessary

TABLE 15.3a International Comparison of the Strength and Improvements of Science Teacher Preparation Between 14 Countries (Part 1: Europe)

		Europe				
		Portugal	France	U.K. (England)	Germany	Finland
Strength or Improvements	Content Knowledge	• Pure knowledge and a passive teaching and learning method	• Good and sufficient CK is needed to have good PCK			
	General Pedagogical Knowledge	• Many disciplines including assessment, but limited hours and credits • Pedagogical performance				• More knowledge and practice in special education • Using learning technologies and PC • Teacher has to work in groups
	Pedagogical Content Knowledge	• Dialogue educational research-teaching practice • Giving practical works in a investigative way	• Difficult acquiring: unanswered to make an explicit link between training and teacher training	• Need understanding nature of science curriculum	• Need different science teaching between academic and non-academic track	• Importance of in-service education for studying 1st year pedagogical course
	Others	• Emerging guidelines from new research fields like neuroeducation • Ongoing teachers professional learning by regular professional dialogue	• Develop skills of observation & analysis by research activity • Teacher trainers' training program and qualification has just been announced by MOE	• Greater school-controlled ITE would return to partnership model • Same-school two-year training is suitable for partnership model	• After following the Bologna declaration, class became increasingly different • Division of teacher education into an academic and a practical stage	• Department self-evaluation & international evaluation every 5 years as quality control • With research-oriented professional culture

TABLE 15.3b International Comparison of the Strength and Improvements of Science Teacher Preparation Between 14 Countries (Part 2: Middle East, North and Latin America)

		Middle East		North and Latin America		
		Turkey	Israel	Canada	United States	Argentina
Strength or Improvements	Content Knowledge			• Consistency between university admission requirements and CK at undergraduate level	• Alternative licensure program: degree relevant, but less major subject study and pedagogical study	
	General Pedagogical Knowledge	• Introducing constructivist approach to instruction		• Country's history as well as its place within the global community		• Engage educational research projects
	Pedagogical Content Knowledge		• Teaching courses are characterized by dichotomy: Different approachs are needed	• Integrated approach is needed with field experience	• Mainly study during final student teaching, its course hour is limited • Some influence from NGSS	• Constructivist and inquiry-based approaches are needed in science teaching and science teacher education level
	Others	• Little employment of secondary science teacher • Educational environment difference between east and west • Sudden educational system change by government	• Shortage of high school chemistry and physics teacher • Students' negative attitude towards school and teaching subjects: pushing towards an assessment-oriented teaching	• Some provinces: 21st century skills appeared in recent revisions • No national regulatory body for education	• Increase teacher shortage, and decrease the pool of teacher applicants • Increase emergency or alternative licensure teachers	• Expand enrollment rates in secondary level, but less than 50% of the population • Common to occupy positions in several schools and to be paid only for teaching time

TABLE 15.3c International Comparison of the Strength and Improvements of Science Teacher Preparation Between 14 Countries (Part 3: Asia and Pacific)

		Asia and Pacific			
		Australia	China	Korea	Japan
Strength or Improvements	Content Knowledge	• A half of lower secondary science teachers study tertiary science as minor	• CK is the main part in exam-oriented tendency	• Portion of CK course is higher than PCK course • Improve competencies related to instruction	• High school science teacher's licence isn't divided, but needs in-depth main subject CK and other subject's basic CK
	General Pedagogical Knowledge	• 5 National priority areas: indigenous education, classroon management, ICT, literacy and numeracy, special educational needs • Reflective teaching	• Using lecture-oriented strategies • Need alternative assessment to make some suitable change for the standards and exam		• Dealing with modern issues: special educational needs, ICT and active learning
	Pedagogical Content Knowledge	• ASTA standard detailed the knowledge of subject teaching • Stem, classroom discourse • Strong & sustained partnership between univ. & schools	• Using lecture-oriented strategies • Need alternative assessment to make some suitable change for the standards and exam	• Recognize the gap between theory and reality at schools through internship program and plan to reduce the gap	• Scientific performance assessment is needed • Planning a new accreditation criteria: PCKs category will be moved from GPK to CK
	Others	• Progression towards national consistency or frameworks • Education research for effective practice, and initiated shifts in educational policy and curriculum framing	• Certification exam is not related to educational practice	• Elementary and secondary school teachers would be trained in the same institution for providing greater academic knowledge and understanding about student	• Additional field experience and service learning are recommended • planning national teacher qualification test (using for qualification and selection)

for the formation of better PCK. However, learning CK has become a passive activity and just for the examinations' preparation. In addition, the consistency between senior secondary school science CK and CK at the university are also required. The problem is the possibility that teacher trainees end up with inadequate knowledge acquisition for their duties, depending on teacher qualifications and licensing system regulations.

- GPK: The subjects relating to basic pedagogy, didactics, and psychology diverge into many branches, but there is a time limitation and we could not mention all of them. The knowledge acquisition and practical abilities formation related to ICT, special educational needs, foreign language, and classroom management are needed for teacher trainees in response to the real classroom situation. In addition, it is important for teacher trainees to do their practice based on the learning theory, to implement a variety of assessment techniques, and to practice the educational research with self-reflection.
- PCK: The importance of learning PCK supported by the trainee's lesson practice under the university-training school partnership is pointed out. And it is also pointed out that it is necessary for science instruction to introduce an investigative viewpoint, inquiry-based practice, modification of lesson designs suitable for the specialty of students, learning theory of constructivism, various instructional and evaluation methods, and STEM education.
- Others: There are indications from the following four viewpoints: the training of teacher educator, institutional partnership, teacher employment, and educational administration. Regarding the teacher educator's professional ability development, it is recommended to do the research activities for improving the skills of lesson observation and lesson analysis. And it's also recommended taking the periodical qualitative evaluation. In regard to the partnership of the teacher training institution, it is requested to strengthen the collaborative relationship in education and research activities between the university and schools, and among teacher training universities. In regard to the situation of teacher employment, the influences of the lack of secondary school science teachers especially in physics, reduction of the teacher employment, and increase of the substitute license holders is pointed out. And it is also suggested that there is a discontinuity between what is necessary for educational practice and what is asked in teacher qualification examination. Regarding the educational administration policy, it is pointed out the influence of domestic difference in educational environment and radical system reform of science teacher preparation.

Figure 16.1 is a schematic diagram of the aspects of the qualitative reform over the initial teacher training for secondary school science. Most points mentioned above are included in it. There are two kinds of competitive selection for teacher applicants, one is a university entrance examination (entrance of ITE), and another is the teacher employment examination (graduation of ITE). The number of applicants and the nature of passers are affected by the the number of passers or competition rate in recent years. In the case that applicants or school teachers are lacking (need for new teachers), the competition rate will be decreased and passers' knowledge level, it relates mainly to the quantity and quality of science knowledge in CK, will be down. But the passers' knowledge of scientific method in CK is limited and not affected by the competition rate, because secondary school science is not emphasize so much on inquiry activities and two examinations mentioned above usually do not measure the applicants' scientific method skills. Then, teacher trainees need to learn PCK in order to be getting more methodological knowledge including the scientific method in CK. It is also necessary for teacher trainees to do inquiry-based activities themselves and teaching practice for pupils should include inquiry-based activities. By implementing lesson study (educational research), teacher trainees are able to get good feedback from their practice with good suggestions by mentor teachers at training schools. The question stated in the figure, *"What kind and how much teacher knowledge is needed for teacher qualified persons?"* is difficult to answer but worthy of our research.

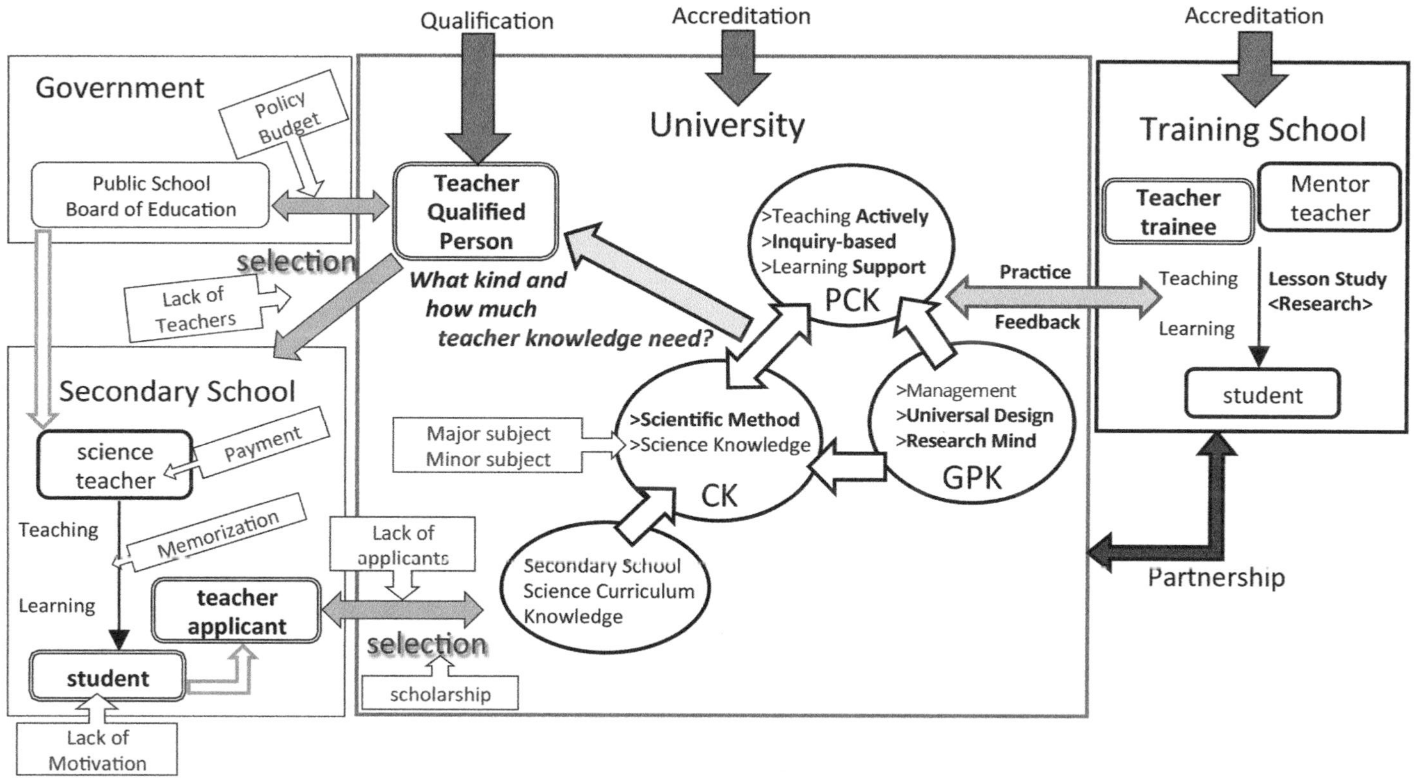

Figure 15.1 Aspect of the qualitative reform over the initial teacher training for secondary school science.

ABOUT THE EDITORS

Jon E. Pedersen is currently the Dean of the College of Education at the University of South Carolina, and author of over one hundred publications: most of which focus on science teaching and/or the incorporation of social issues into the extant curriculum. His interest in the application of science knowledge as it pertains to relevant social issues has been greatly influenced by his foundation in the agricultural sciences and his early years growing up on a farm. He has also published 16 books and has been primary investigator and co-primary investigator of numerous grants (over 8 million in funded activities) and supported projects on science curricula development, science in-service education, middle level education, and international education totaling well over four million dollars. Over the years, Dr. Pedersen has also worked in more than a dozen different countries around the world. Most notably is his work in Bolivia, South America assisting teachers from the rural areas of the country in understanding how science relates to the lives of the youth and families in these communities. Dr. Pedersen is very active in several professional organizations including: Nebraska Association for Science Teachers (NATS; past president), National Science Teachers Association (NSTA; board member), National Association for Research in Science Teaching, Association for Science Teacher Education (past president), American Association for Teacher Education (AACTE), Organization of Institutional Affiliates (OIA), and the American Educational Research Association (AERA).

Model Science Teacher Preparation Programs, pages 327–328

Tetsuo Isozaki is a professor of Graduate School of Education, Hiroshima University. He has researched and taught about history of science education for more than twenty years. His research focuses on history of science education with comparative study, and science teacher education. Dr. Isozaki is active in several professional organizations including: East-Asian Association for Science Education (EMs), European Science Education Research Association, Society of Japan Science Teaching (a member of Board of Directors), Japan Society for Science Education (Councilor), and Japan Society of Earth Science Education (a member of Standing Committee).

Toshihde Hirano is a professor of science education at Aichi University of Education, Japan. He received his PhD in Curriculum and Instruction Sciences (Science Education Major) from Hiroshima University, Japan. He began his career in primary and secondary school teacher education and training at the Faculty of Education of Shimane University in 2000. He is interested in the influences of the structure of curricula and group communications on learners' concept construction in science, and in the comparative study of science education and teacher education with East Asian countries and the United States. Dr. Hirano is currently a board member in the Society of Japan Science Teaching (SJST).

Printed and bound by CPI Group (UK) Ltd, Croydon, CR0 4YY

15/09/2026

14962337-0001